Editor
Mary S. Jones, M.A.

Illustrator
Mark Mason

Cover Artist
Brenda DiAntonis

Managing Editor
Karen J. Goldfluss, M.S. Ed.

Creative Director
Karen J. Goldfluss, M.S. Ed.

Art Production Manager
Kevin Barnes

Art Coordinator
Renée Christine Yates

Imaging
Denise Thomas
Nathan Rivera

Publisher
Mary D. Smith, M.S. Ed.

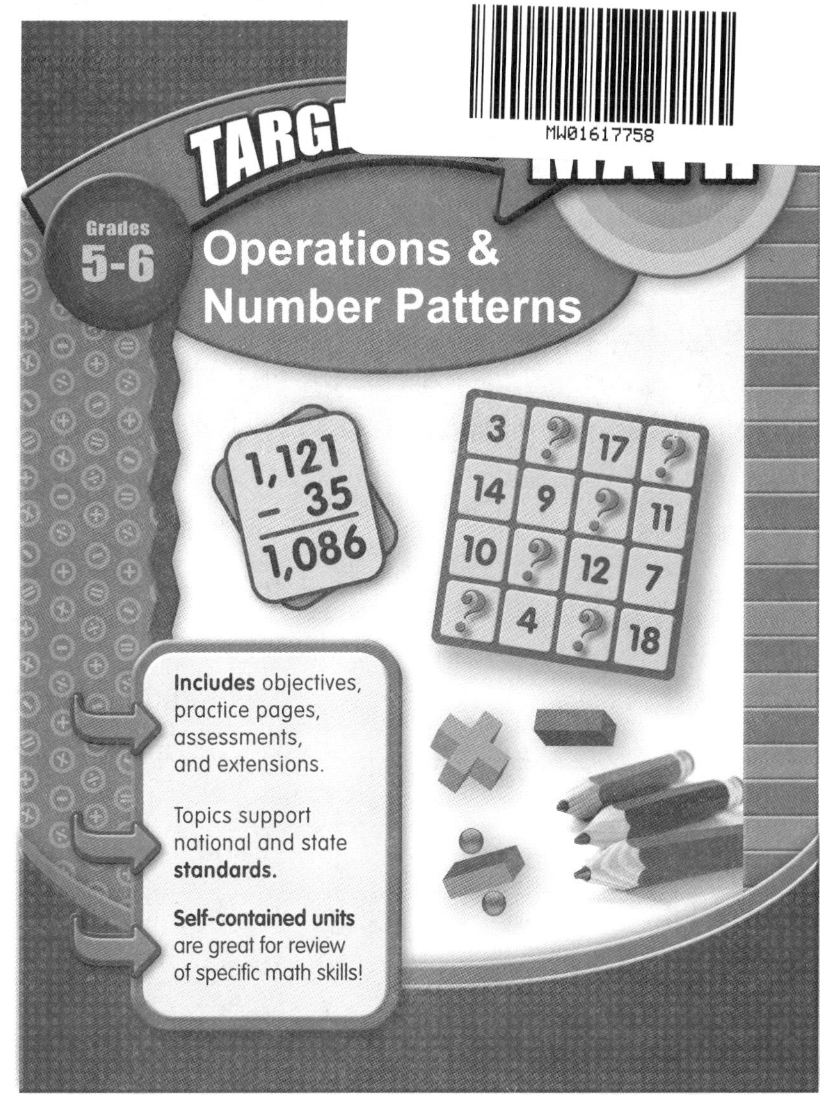

Authors

Richard Glasson, Gloria Harris,
Garda Turner, and Kerry Walker

(Revised and rewritten by Teacher Created Resources, Inc.)

Teacher Created Resources, Inc.
6421 Industry Way
Westminster, CA 92683
www.teachercreated.com
ISBN-13: 978-1-4206-8998-3

Made in U.S.A.

Table of Contents

Table of Contents

Introduction

Targeting Math

The *Targeting Math* series is a comprehensive classroom resource. It has been developed so that teachers can find activities and reproducible pages for all areas of the primary math curriculum.

About This Series

The twelve books in the series cover all aspects of the math curriculum in an easy-to-read format. Each level—grades 1 and 2, grades 3 and 4, and grades 5 and 6—has four books: Numeration and Fractions; Operations and Number Patterns; Geometry, Chance and Data; and Measurement. Each topic in a book is covered by one or more units that are progressive in level. The teacher is able to find resources for all his or her students, whatever their ability. This enables the teacher to differentiate for different ability groups. It also provides an easy way to find worksheets at different levels for remediation and extension.

About This Book

Targeting Math: Operations and Number Patterns (Grades 5 and 6) contains the following topics: addition, subtraction, multiplication, division, number facts, and number patterns. Every topic contains two complete units of work. (See Table of Contents for specific skills.)

About Each Unit

Each unit is complete in itself. It begins with a list of objectives, resources needed, mathematical language used, and a description of each reproducible student page. This is followed by suggested student activities to reinforce learning. The reproducible pages cover different aspects of the topic in a progressive nature and all answers are included. Every unit includes an assessment page. Each assessment page is an important resource in itself as teachers can use each one to find out what their students know about a new topic. They can also be used for assessing specific objectives when clear feedback is needed.

About the Skills Index

A Skills Index is provided at the end of the book. It lists specific objectives for the student pages of each unit in the book.

ADDITION

These units contain exercises in horizontal and vertical algorithms, adding quantities, problem solving, completing addition patterns, money, and decimals.

Skills are also practiced by estimating answers with a greater degree of accuracy, finding averages, calculating angle sums, and working out perimeters.

Students are encouraged to use appropriate mathematical language, differentiate between sensible estimates and exact answers, and draw on numerous strategies when working mentally.

There is an activity page that contains a game and two addition tasks solved by trial and error. Two assessment pages are included.

UNDERSTANDING ADDITION

Unit 1

Horizontal addition
Estimation
Averages
Missing digits
Decimals

Objectives

- *make a suitable choice of operation involving whole numbers*
- *mentally add two-digit numbers*
- *estimate sums by rounding*
- *use a variety of ways, when prompted, to check work*
- *use approximate additions to check that the decimal places in a calculation are reasonable*
- *use a variety of informal strategies to approximate, calculate, and represent solutions to addition problems*
- *select and use appropriate operations to deal with situations which may involve large whole numbers, money, and decimal fractions*
- *use number skills involving whole numbers to solve problems*
- *select appropriate technology and use it to help carry out mathematical investigations*
- *use understood written methods to add*

Language

column, vertically, horizontally, estimate, average, mean, addition, decimal, calculations, sum, total

Materials/Resources

dice, calculators, scratch paper, 0–9 spinners or polyhedral dice

Contents of Student Pages

* *Materials needed for each reproducible student page*

Page 8 Correct Columns
adding to 99,999; horizontal and vertical forms of addition; addition games
* *dice, scratch paper*

Page 9 Estimating
speed drill; estimating then calculating; addition wheels
* *calculators*

Page 10 Averages
finding the mean of a set of scores; problem solving

Page 11 Missing Digits
working backwards to fill in missing digits; < and >; true or false; addition game
* *scratch paper*

Page 12 Decimals
adding decimals; horizontal and vertical forms of addition; problem solving

Page 13 Assessment

Page 14 Reaching 500 Activity
finding addends
* *scratch paper, 0–9 spinners or polyhedral dice*

Remember

Encoutrage students to:

- ❑ *add from the top.*
- ❑ *record carry numbers.*
- ❑ *be aware of placing numerals in the correct columns.*

Additional Activities

- ❑ *Students bring in magazines and catalogues. They go on a shopping spree and total their purchases.*
- ❑ *Keep a record of temperatures for a week/month and find the mean temperature.*
- ❑ *Find the class average:*
  - *age*
  - *spelling test scores*
  - *weight, etc.*
- ❑ *Use numeral cards in many ways to practice addition.*

Answers

Page 8 Correct Columns

1. a. 16,847
   b. 16,414
   c. 9,801
   d. 30,723
2. a. 12,241
   b. 1,140
   c. 14,695
   d. 24,088
   e. 7,233
   f. 12,994
   g. 5,636
   h. 10,634
3. Answers will vary.

Page 9 Estimating

1.

12	16	21	17	13	39	18	14	30	15
7	11	16	12	8	34	13	9	25	10
10	14	19	15	11	37	16	12	28	13
8	12	17	13	9	35	14	10	26	11

2. a. 2,196
   b. 879
   c. 1,636
   d. 997
   e. 2,222
   f. 444
   g. 2,106
3. a. 56, 89, 123, 48, 142, 101, 54, 78
   b. 67, 110, 125, 100, 208, 122, 86, 159

Page 10 Averages

1. 56
2. 47
3. 21.4°F
4. 11 years 5 months
5. 1 hr. 17 min.
6. 17.4

Page 11 Missing Digits

1. a. 3, 5
   b. 3, 1, 9
   c. 8, 7, 3
   d. 3, 8
   e. 8, 0
   f. 8, 9
   g. 7, 1
   h. 9, 2
2. a. F f. F
   b. F g. T
   c. T h. F
   d. F i. T
   e. F j. T
3. Answers may vary.

Page 12 Decimals

1. a. 28.1
   b. 48.2
   c. 112.7
   d. 89.3
   e. 74.4
   f. 269.03
   g. 119.40
   h. 6.729
   i. 1.943
   j. 107.403
2. a. 7.9
   b. 2.14
   c. 12
   d. 14.1
   e. 12
   f. 28
   g. 20.42
   h. 37
3. a. $28.75
   b. 9.11 ft.

Page 13 Assessment

1. a. 1,787 c. 260,437
   b. 23,623 d. $19,434.20
2. a. 230.19
   b. 33,392
   c. 83.119
3. a. 30, 25, 41, 53, 32, 69, 64, 148
   b. 3.9, 10.6, 5.4, 6.1, 2.3, 9.8, 13.2, 11
4. 9
5. a. 5, 3, 1 c. 4, 3
   b. 8, 5, 7 d. 9, 1, 9

Page 14 Reaching 500 Activity

1. 723 + 2,451 + 905 + 382 + 87 + 4,094 + 868 + 489
2. 17,206 + 38,057 + 8,932 + 7,228 + 5,059 + 23,517

Name **Date**

1\. **a.**

7,384
691
8,725
+ 47

b.

75
6,914
378
+ 9,047

c.

3,089
5,974
682
+ 56

d.

8,300
9,849
7,028
+ 5,546

2\. Set these out vertically to solve. Use the space at the right and/or scratch paper.

a. 5,328 + 796 + 23 + 6,094

b. 51 + 729 + 6 + 354

c. 216 + 9,891 + 3,723 + 865

d. 6,182 + 8,504 + 75 + 9,327

e. 5,639 + 17 + 1,024 + 553

f. 94 + 380 + 5,619 + 6,842 + 59

g. 1,652 + 2,384 + 917 + 38 + 645

h. 8 + 2,098 + 69 + 811 + 7,648

3\. Play this game with two other students. Take turns to roll four dice. Record the largest number you can. After five turns, total your scores. The one with the highest score wins. Play it again writing down the lowest scores you can. The one with the lowest total is the winner.

Highest Scores

Player 1	Player 2	Player 3

Lowest Scores

Player 1	Player 2	Player 3

Name **Date**

1. Complete the following table without using a calculator.

+	3	7	12	8	4	30	9	5	21	6
9										
4										
7										
5										

2. Estimate the answers to the nearest hundred. Then use a calculator to see how close you were.

a. 824 + 758 + 614 = Est. ____________ Ans. ____________

b. 87 + 224 + 568 = Est. ____________ Ans. ____________

c. 946 + 581 + 109 = Est. ____________ Ans. ____________

d. 337 + 43 + 617 = Est. ____________ Ans. ____________

e. 774 + 458 + 990 = Est. ____________ Ans. ____________

f. 13 + 29 + 402 = Est. ____________ Ans. ____________

g. 551 + 739 + 816 = Est. ____________ Ans. ____________

3. Use your brainpower to work these mentally.

a.

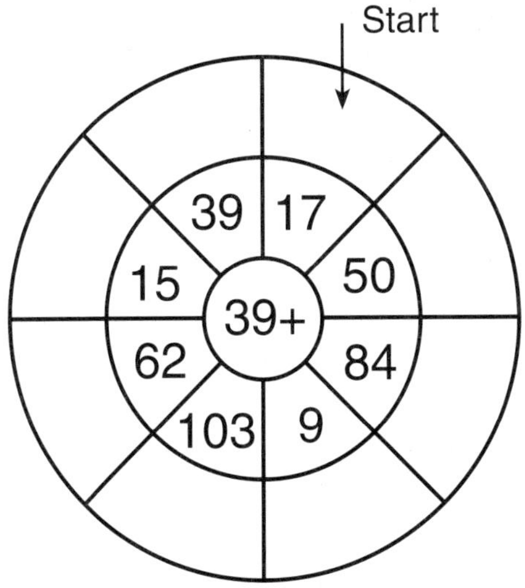

b.

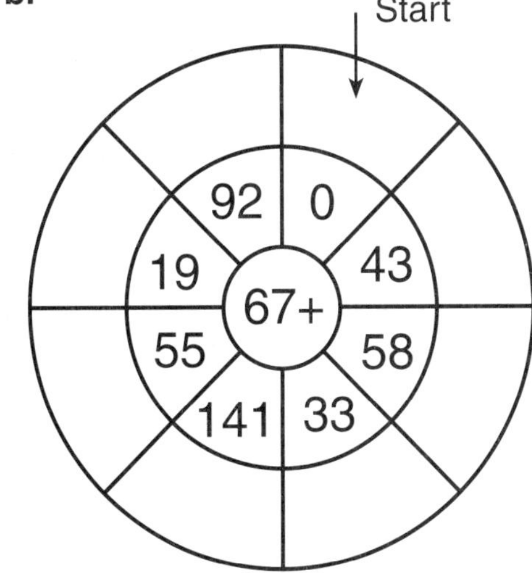

Name **Date**

To find the **average** of a set of scores, add all the scores and divide the answer by the total number of scores.

Example: The average of 10, 14, 17, 21, and 33 is:

$10 + 14 + 17 + 21 + 33 = 95$

$\text{Average} = 95 \div 5$

$= 19$

Another word for average is **mean**.

Write a number sentence to find the answer for each problem.

1. Find the average of these scores: 72, 69, 58, 31, 84, 22.

 ____

2. Sam played in several cricket games. His batting scores were 67, 3, 29, 48, 107, 19, 56. What was his batting average?

 ____

3. For one cold, winter school week, the daily temperatures were 21°F, 25°F, 19°F, 22°F, and 20°F. What was the mean daily temperature?

 ____

4. Jo is 11 years 1 month old, Angela is 11 years 6 months old, Megan is 10 years 7 months old, and Kate is 12 years 6 months old. What is the average age of these friends?

 ____

5. Every morning Richard goes for a jog. On Monday he jogged for 1 hour 17 min., Tuesday 1 hour 11 min., Wednesday 1 hour 15 min., Thursday 59 min., and Friday 1 hour 43 min. What was the average length of time he jogged each day?

 ____

6. On her 10 spelling tests, Liza's scores were 17, 20, 15, 19, 20, 18, 16, 11, 18, 20. What was her mean score?

 ____

Name **Date**

1. Put numerals in the squares to make these additions correct.

a. $\begin{array}{r} 6\ 9 \\ 5\ \square \\ +\ 3\ 6 \\ \hline 1\ \square\ 8 \end{array}$

b. $\begin{array}{r} \square\ 7 \\ 4\ 9 \\ +\ 8\ 3 \\ \hline \square\ 6\ \square \end{array}$

c. $\begin{array}{r} 4\ \square\ 7 \\ 5\ 9\ 1 \\ +\ \square\ 6\ 5 \\ \hline 1\ 8\ 4\ \square \end{array}$

d. $\begin{array}{r} 3\ 9 \\ \square\ 6\ 3 \\ +\ 5\ 7\ \square \\ \hline 9\ 8\ 0 \end{array}$

e. $3\square + 43 + 27 = 1\square 8$

f. $\square 8 + 4\square + 42 = 179$

g. $0.5 + \square.2 + 9.3 = \square 7$

h. $2\square.2 + 3.17 + 4.\square = 36.57$

2. True or false?

a. $17 + 64 < 38 + 39$ ____________

b. $7^2 = 16 + 23$ ____________

c. $49 + 216 > 196 + 59$ ____________

d. $68 + 84 = 12^2$ ____________

e. 8 dozen $< 4^2 + 80$ ____________

f. $296 + 385 < 396 + 285$ ____________

g. $7 + 9 + 6 + 8 + 11 < 7^2$ ____________

h. $500 > 215 + 175 + 110$ ____________

i. $323 = 86 + 79 + 95 + 63$ ____________

j. $6^2 + 7^2 + 8^2 > 11^2 + 5^2$ ____________

3. Find the highest score.

- You must start on a number on the top row.
- You must finish on a number on the bottom row.
- You can make only 20 moves.
- Each move may only be vertical or horizontal.
- Add the numbers as you move. You will need scratch paper to record your tries.

What is your highest score?

____

2	7	9	6	1	4	8	2
8	3	5	7	2	5	0	4
7	0	8	2	9	6	3	8
2	3	5	0	1	3	7	5
5	1	4	6	9	4	0	2
6	8	2	7	3	9	4	6
1	9	0	5	4	0	3	8
2	7	3	8	1	6	9	1

Name **Date**

When **adding decimals**, remember to:
- always estimate first.
- keep the decimal points underneath each other.
- put the decimal point in the answer.

1. Estimate in whole numbers before working out the correct answer.

a.	b.	c.	d.	e.
6.9	18.2	68.7	0.6	48
7.2	7.5	9	36.2	2.3
5.3	13.1	23.4	49.5	19.1
+ 8.7	+ 9.4	+ 11.6	+ 3	+ 5
Est.	Est.	Est.	Est.	Est.

f.	g.	h.	i.	j.
96.23	88.02	3.756	0.3	38.614
80.05	9.6	0.82	0.959	2.95
75.91	18	1.953	0.68	0.039
+ 16.84	+ 3.78	+ 0.2	+ 0.004	+ 65.8
Est.	Est.	Est.	Est.	Est.

2. Work these out mentally.

a. 2.4 + 3.9 + 1.6 = ____________ **b.** 0.97 + 0.23 + 0.94 = ____________

c. 0.7 + 8.5 + 2.8 = ____________ **d.** 8.1 + 0.9 + 3.6 + 1.5 = ____________

e. 9 + 2.6 + 0.4 = ____________ **f.** 3 + 6.8 + 7.2 + 11 = ____________

g. 16 + 0.55 + 3.87 = ____________ **h.** 1.25 + 26.2 + 9.55 = ____________

3. **a.** Debby had fourteen 25-cent coins, forty-nine 5-cent coins, sixty-three 10-cent coins, and thirty-three 50-cent coins. How much did she have altogether?

____

b. To make a model, John must have five pieces of rod measuring 1.63 ft., 0.95 ft., 2.1 ft., 1.56 ft., and 2.87 ft. What was the total length of rod he needed?

____

Name   **Date**

1. Estimate before adding.

a.	**b.**	**c.**	**d.**
674	2,714	26,818	\$2,561.45
392	8,309	60,251	\$7,394.68
586	4,573	98,320	\$6,027.32
+ 135	+ 8,027	+ 75,048	+ \$3,450.75
______	______	______	______
Est. ______	Est. ______	Est. ______	Est. ______

2. Set these out vertically in the space provided before adding.

a. 96.84 + 32.96 + 83.05 + 17.34 = ____________

b. 27,615 + 98 + 385 + 5,294 = ______________

c. 8.709 + 38.3 + 0.51 + 27 + 8.6 = ____________

3. Complete these using mental calculations.

a.

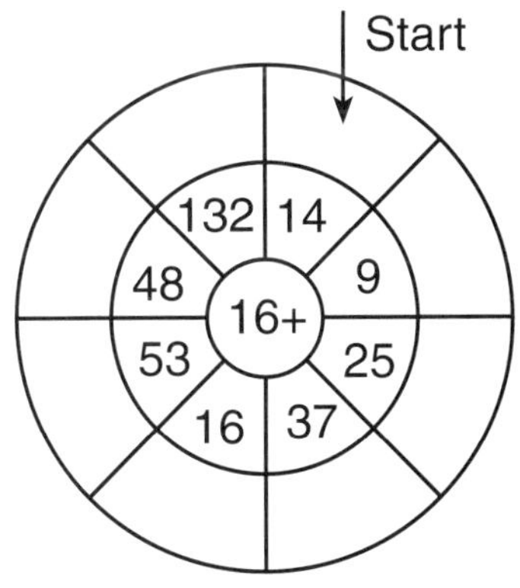

b.

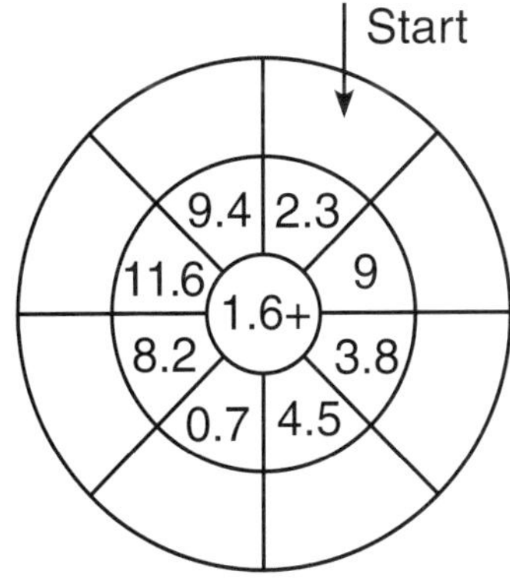

4. What is the mean of 12, 8, 9, 5, 7, 16, 6? ________________________________

5. Fill in the missing numerals.

a.

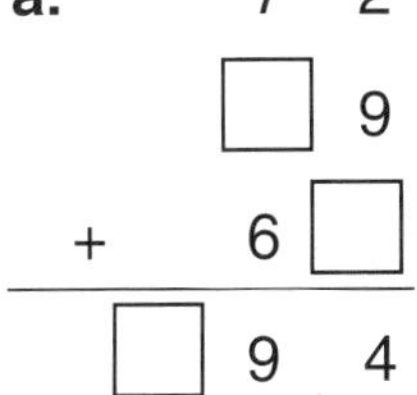

b.

6 ☐ ☐
4 2 5
+ ☐ 8 7
1, 8 9 7

c. 0.8 + ☐.6 + 2.☐ = 7.7

d. 1☐.56 + ☐7.93 = 37.4☐

Name	**Date**

Number of players: 2–4

Equipment: scratch paper for each player

two 0–9 spinners (or polyhedral dice)

Object: to be the first player to reach 500

How to play: Spin both spinners, make a two-digit number, and mentally add it to your total. Write down your new total. If you make a mistake (the other players will tell you), you must subtract it from your score.

Variation: To make the game more difficult, the winner must reach 500 exactly.

Players can subtract the two-digit numbers, as well as add them, to help reach the target.

---

1. Circle the eight numbers that total 9,999.

723 + 2,451 + 819 + 905 + 67 + 215 + 382 + 87 +

4,013 + 4,094 + 828 + 868 + 2,118 + 489 = 9,999

2. Circle the six numbers that total 99,999.

12,053 + 17,206 + 38,057 + 5,914 + 11,320 + 8,932 +

564 + 7,228 + 5,059 + 24,193 + 23,517 = 99,999

WORKING WITH ADDITION

Unit 2

Large numbers
Number types
Quantities
Money
Problem solving
Working mentally

Objectives

- *make a suitable choice of operations involving whole numbers*
- *mentally add two-digit numbers*
- *use understood written methods to add*
- *use a variety of informal strategies to approximate, calcualate, and represent solutions to addition problems*
- *select and use appropriate sequences of operations to solve problems*

Language

odd, even, prime, composite, addition, sum, total, average, vertically, horizontally, column, quantities, perimeter

Materials/Resources

scratch paper

Contents of Student Pages

* *Materials needed for each reproducible student page*

Page 17 Number Types
odd, even, prime, composite; sum, total; mental grid
* *scratch paper*

Page 18 Adding to 99,999
money; missing numbers; vertical and horizontal forms
* *scratch paper*

Page 19 Mystery Totals
largest total, smallest total; angle sum; average age

Page 20 Adding Quantities
time, length, weight

Page 21 Working Mentally
perimeters; short cuts; patterns; problem solving

Page 22 Assessment

Remember

- ❑ *Discuss the benefit of always estimating.*
- ❑ *Practice converting quantities, e.g., seconds to minutes to hours.*

Additional Activities

- *Use maps to work out distances of long trips.*
- *Use timetables to work out time taken for long excursions.*
- *Students work in groups of three to make an addition game.*

Answers

Page 17 Number Types

1. a. A whole number that ends in 1, 3, 5, 7, or 9.
   b. A whole number that is divisible by 2.
   c. A number that has exactly two positive factors, itself and 1.
   d. A number that has more than two factors.
2. a. 1,070
   b. 225
   c. 38.5
   d. 340
3.

33	27	62	86	39	14	100	78	95	51
28	22	57	81	34	9	95	73	90	46
31	25	60	84	37	12	98	76	93	49
29	23	58	82	35	10	96	74	91	47
32	26	61	85	38	13	99	77	94	50

4. 471, 530, 551, 618, 714, 789, 807

Page 18 Adding to 99,999

1. a. 248,203
   b. 194,058
   c. 141,143
   d. 1,644,750
   e. $2,621.12
   f. $1,690.03
   g. 2,861.36 ft.
   h. 115.676 ft.
2. a. 18
   b. 23
   c. 35
   d. 57
   e. 155
   f. 49
3. a. 108,911
   b. 109,402
   c. 183,916
   d. 143,978
   e. 121,860

Page 19 Mystery Totals

1. Check individual work. Possible answers:
   a. 98,765,432 + 1 + 0 = 98,765,433
   b. 1,049 + 258 + 367 = 1,674
2. Check individual work. Possible answers:
   a. 9,876,543 + 2 + 1 + 0 = 9,876,546
   b. 109 + 238 + 46 + 57 = 450
3. a. 720°
   b. 1,260°
4. Answers will vary.

Page 20 Adding Quantities

1. a. 3, 8, 1, 6
   b. 7, 8, 0, 2
   c. 5, 3, 8, 4
   d. 7, 5, 8, 9
2. a. 12 hr. 7 min.
   b. 54 yr. 8 months
   c. 48 min. 23 sec.
   d. 81 L 188 mL
   e. 913 m 92 cm
   f. 156 kg 590 g
   g. 30 hr. 27 min. 45 sec.
   h. 104 days 10 hr. 4 min.
   i. 18 m 26 cm 4 mm
3. Check individual work.

Page 21 Working Mentally

1. a. 66 ft.
   b. 32.2 cm
2. a. 20
   b. 29
   c. 62
   d. 89
   e. 175
   f. 120
   g. 201
   h. 220
3. a. 188, 207, 226, 245, 264, 283
   b. 230, 247, 264, 281, 298, 315
   c. 453, 478, 503, 528, 553, 578
   d. 153, 199, 245, 291, 337, 383
   e. 479, 602, 725, 848, 971, 1,094
4. a. $12,721.75
   b. 18

Page 22 Assessment

1. a. F
   b. T
   c. T
   d. T
2. a. 13 hr. 41 min.
   b. 588 L 346 mL
   c. 17 hr. 10 min. 56 sec.
3. a. 301, 308, 315, 322, 329
   b. 194, 210, 226, 242, 258
   c. 21.6, 25, 28.4, 31.8, 35.2
4. 77.8 cm or 778 mm
5. 540°
6. a. 15,840 ft. or 3 miles
   b. 2,640 ft. or 0.5 miles

Name Date

1. a. What is an odd number?

____

b. What is an even number?

____

c. What is a prime number?

____

d. What is a composite number?

____

2. Do your work on scratch paper.
   a. Total the next 10 even numbers after 96. ________
   b. What is the sum of the first 15 odd numbers? ________
   c. What is the average of the first 10 square numbers? ________
   d. What is the total of the next eight prime numbers after 24? ________

3. Add these mentally. Time yourself. Can you complete the table in under four minutes?

+	24	18	53	77	30	5	91	69	86	42
9										
4										
7										
5										
8										

Time ____________ Score ____________

4. Work mentally.

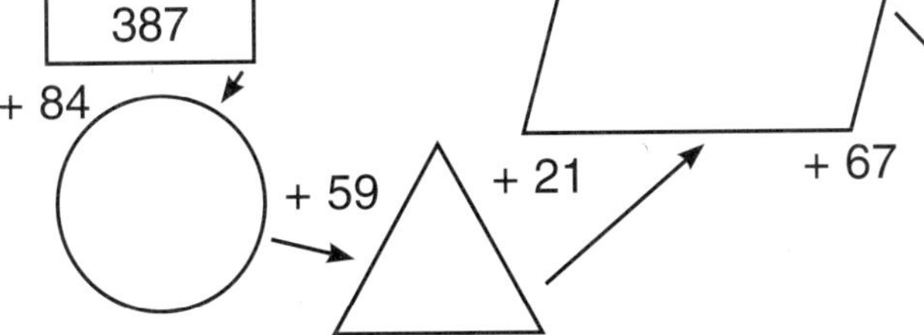

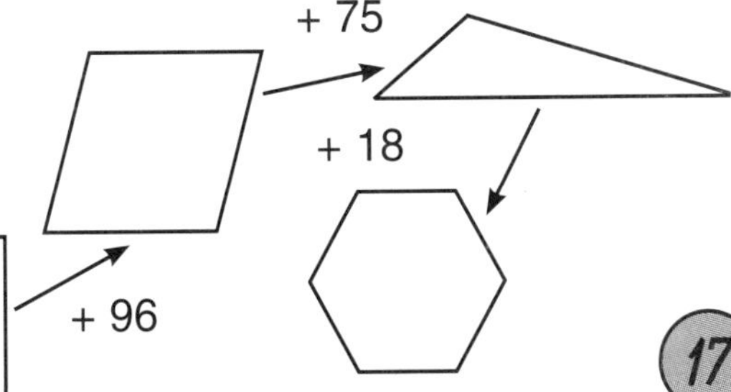

Name **Date**

1.

a.	b.	c.	d.
72,943	69,714	77,915	993,265
86,507	8,355	207	2,078
20,391	88,032	5,614	41,309
38,477	27,360	48,321	526,672
+ 29,885	+ 597	+ 9,086	+ 81,426

e.	f.	g.	h.
$236.15	$760.00	618.42 ft.	63.251 ft.
$964.03	$93.84	923.75 ft.	9.472 ft.
$872.66	$6.57	861.93 ft.	28.035 ft.
+ $548.28	+ $829.62	+ 457.26 ft.	+ 14.918 ft.

2. Fill in the missing numbers.

a. 17 + 16 + ☐ + 5 = 56

b. ☐ + 9 + 64 + 17 = 113

c. (6 x 5) + ☐ + 72 = 137

d. 219 = 73 + ☐ + 89

e. (7 x 14) + (19 x 3) = ☐

f. 24 + 96 = 71 + ☐

3. Set these out vertically on scratch paper or in the space below before adding. Remember to put numbers in the right columns.

a. 96 + 23,148 + 719 + 5,642 + 79,306

b. 19,030 + 82,654 + 78 + 636 + 2,375 + 4,621 + 8

c. 38 + 561 + 9,716 + 92,547 + 80,290 + 764

d. 26,500 + 4,200 + 97,916 + 83 + 947 + 14,238 + 94

e. 479 + 31,516 + 25 + 9 + 51,889 + 280 + 37,662

Name **Date**

1. You have 13 cards:
   - 10 numeral cards 0–9
   - two [+] cards
   - one [=] card

   a. What is the largest total you can make using all the cards?

   Number sentence ______________________ Total ____________

   b. What is the smallest total you can make using all the cards?

   Number sentence ______________________ Total ____________

2. Repeat the above exercise using three [+] cards.

   a. What is the largest total you can make using all the cards?

   Number sentence ______________________ Total ____________

   b. What is the smallest total you can make using all the cards?

   Number sentence ______________________ Total ____________

3. Find the angle sum of the angles of these shapes.

a.

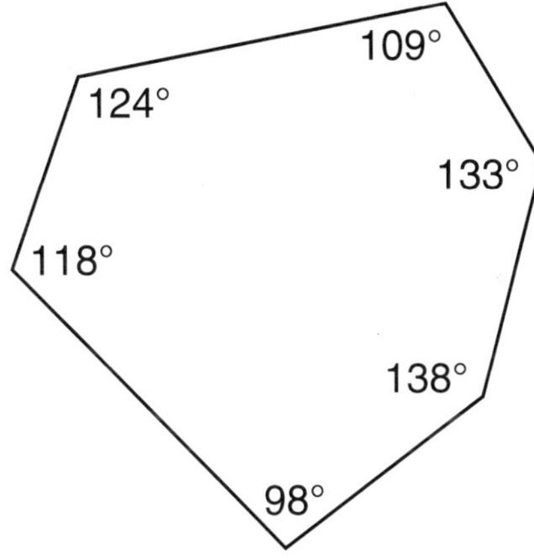

The angle sum of a hexagon is ________ .

b.

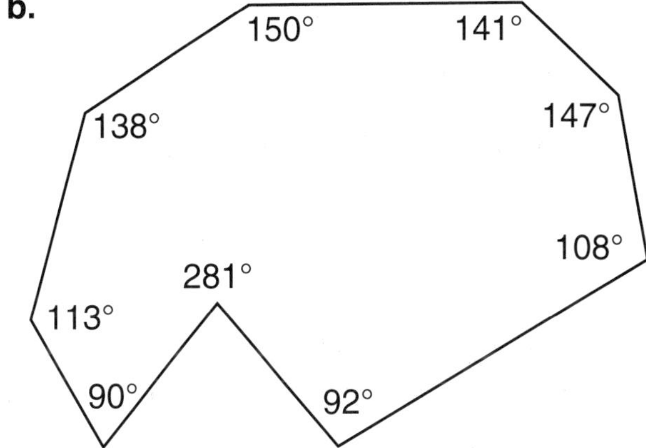

The angle sum of a nonagon is ________ .

4. Use the table to show your work.

   a. What are the ages in months of the students in your class?

   b. What is the total of these ages?

   ____

   c. What is the average age in months? ____________

   d. What is the average age in years?

   ____

Ages in Months		Average Age

Name **Date**

1. Fill in the missing digits.

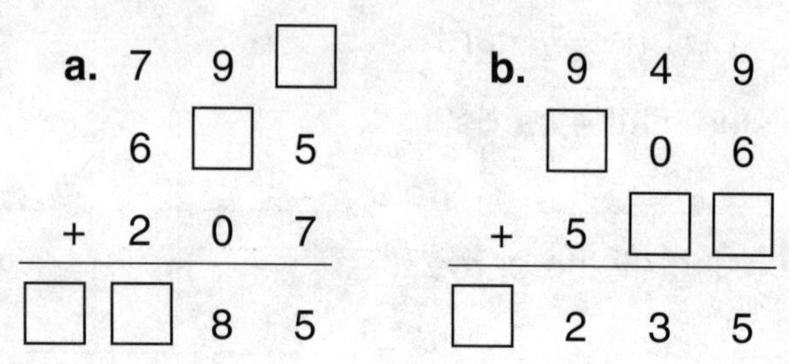

a.	b.	c.	d.
7 9 □	9 4 9	1 □ □ □	5 □ □ 2
6 □ 5	□ 0 6	2 9 4	3 3 9 1
+ 2 0 7	+ 5 □ □	+ 2 7 5 6	+ 1 4 □
□ □ 8 5	□ 2 3 5	□ 5 8 8	□ 2 9 1

2. Be careful when adding quantities. You may have to do some conversions.

Example:

	7 yr.	9 months
+	8 yr.	7 months
	15 yr.	16 months
+	1 yr.	–12 months
	16 yr.	4 months

a.
7 hr. 39 min.
+ 4 hr. 28 min.

b.
14 yr. 9 months
+ 39 yr. 11 months

c.
28 min. 37 sec.
+ 19 min. 46 sec.

d.
27 L 236 mL
18 L 194 mL
+ 35 L 758 mL

e.
127 m 94 cm
576 m 25 cm
+ 209 m 73 cm

f.
96 kg 517 g
42 kg 380 g
+ 17 kg 693 g

g.
4 hrs. 27 min. 35 sec.
18 hrs. 48 min. 41 sec.
+ 7 hrs. 11 min. 29 sec.

h.
37 days 19 hr. 33 min.
49 days 15 hr. 46 min.
+ 16 days 22 hr. 45 min.

i.
8 m 73 cm 9 mm
3 m 58 cm 7 mm
+ 5 m 93 cm 8 mm

3. Make up three problems of your own. Ask a friend to work out the answers. Check that your friend is correct.

a. **b.** **c.**

Name **Date**

1. Find the perimeter of each shape. Be careful with the measurements!

a.

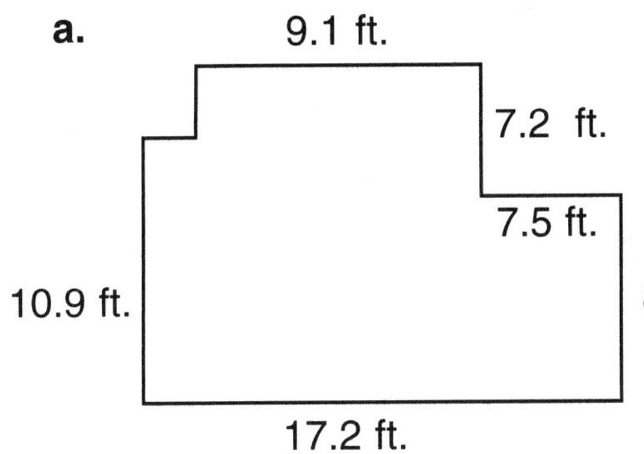

P = ____________________

b.

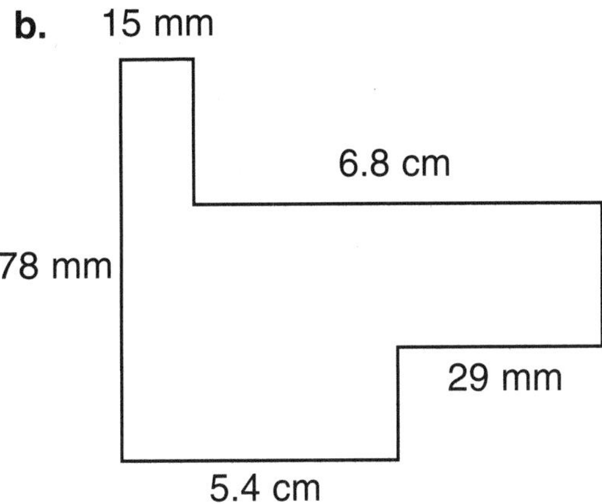

P = ____________________

2. Add these mentally. Look for shortcuts.

a. 6 + 9 + 4 + 1 = ____________

b. 8 + 9 + 2 + 7 + 3 = ____________

c. 17 + 15 + 13 + 5 + 12 = ____________

d. 39 + 26 + 24 = ____________

e. 67 + 75 + 33 = ____________

f. 45 + 28 + 15 + 32 = ____________

g. 70 + 63 + 38 + 30 = ____________

h. 91 + 89 + 24 + 16 = ____________

3. Follow the instructions to complete the patterns.

a. Add 19 | 169 ________ ________ ________ ________ ________ ________

b. Add 17 | 213 ________ ________ ________ ________ ________ ________

c. Add 25 | 428 ________ ________ ________ ________ ________ ________

d. Add 46 | 107 ________ ________ ________ ________ ________ ________

e. Add 123 | 356 ________ ________ ________ ________ ________ ________

4. A school was raising money to buy some musical instruments. In March they collected $1,217.65, in April $3,614.55, in May $5,094.70, and in June $2,794.85.

a. How much did they raise altogether?____________________

b. If instruments average $700 each, how many could they buy?____________________

Name **Date**

1. True or false?

   a. $69 + 54 < 72 + 41$ ________ **b.** $83 + 97 > 94 + 76$ ________

   c. $126 + 593 = 387 + 332$ ________ **d.** $847 + 689 < 794 + 803$ ________

2. **a.**
   5 hr. 53 min.
   \+ 7 hr. 48 min.

   b.
   283 L 517 mL
   \+ 304 L 829 mL

   c.
   2 hr. 41 min. 15 sec.
   4 hr. 38 min. 59 sec.
   \+ 9 hr. 50 min. 42 sec.

3. Follow the instructions to complete these patterns. Work mentally.

   a. Add 7 | 294 ________ ________ ________ ________ ________

   b. Add 16 | 178 ________ ________ ________ ________ ________

   c. Add 3.4 | 18.2 ________ ________ ________ ________ ________

4. Find the perimeter. Watch the units!

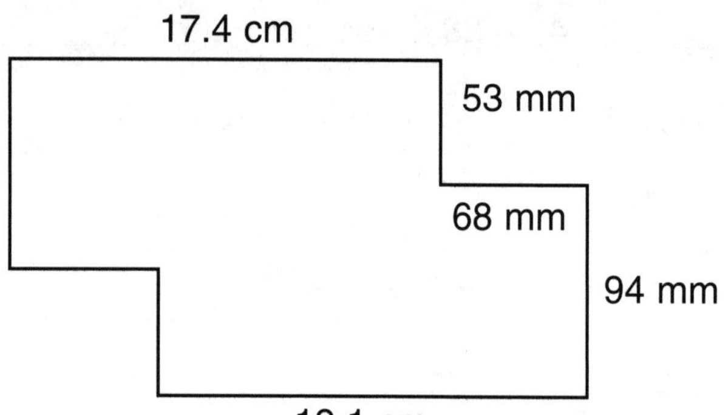

P = ________________

5. Find the angle sum.

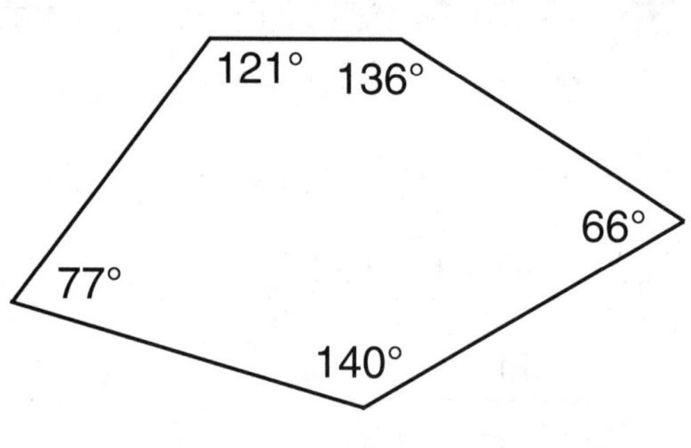

Angle sum = ________________

6. For her morning exercises, Sally walked the following distances:

Monday	2,865 ft.	Tuesday	2,394 ft.	She rested on Sunday.
Wednesday	2,617 ft.	Thursday	2,561 ft.	
Friday	2,218 ft.	Saturday	3,185 ft.	

   a. How far did she walk in the week? ________________

   b. What was her average daily walk? ________________

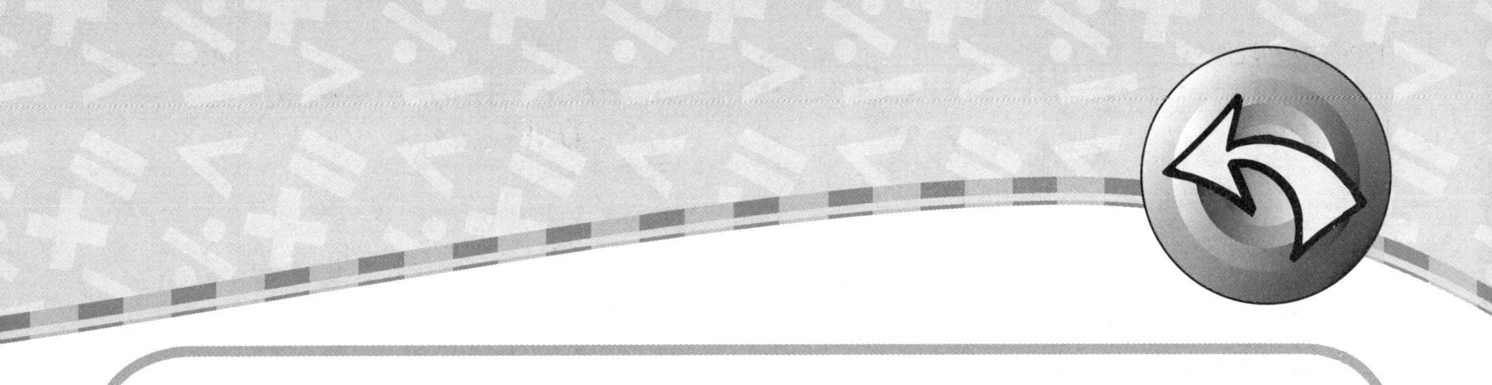

SUBTRACTION

These are two subtraction units that contain exercises in subtraction up to six figures, with and without regrouping. Both vertical and horizontal algorithms are used. Whole numbers, decimal fractions, money, and measurements are included in the exercises.

Skills are further practiced by solving problems, completing subtraction wheels, and breaking codes. There are games to play to reinforce strategies. Estimation is encouraged as is the appropriate setting out and the use of correct mathematical language.

There are two assessment pages and an activity page where students explore magic subtraction squares.

UNDERSTANDING SUBTRACTION

Unit 1

Subtraction to 999
Trading
Problem solving
Code breaker
Money
Decimals

Objectives

- use a calculator efficiently for operating on numbers
- use understood written methods to subtract whole numbers, money, and measurements
- use subtraction skills involving whole numbers to solve problems
- select appropriate strategies to approximate and calculate solutions to problems involving whole numbers, money, and decimal fractions
- appreciate how mathematics is used in a range of aspects of society
- select and use appropriate problem-solving strategies to complete investigations
- select appropriate technology and use it to help carry out investigations

Language

minus, subtract, less, difference, how many left, how many more/less, take away, rename, regroup

Materials/Resources

calculators, scratch paper

Contents of Student Pages

** Materials needed for each reproducible student page*

Page 26 Subtraction to 999
subtracting then checking by addition; working backwards to fill in the missing digits; cross number puzzle

Page 27 Number Smart
round to the nearest 10 and 100; estimating by rounding to nearest 10 or 100; number wheels
** calculators*

Page 28 Trade Away
horizontal and vertical forms of subtraction; subtraction with zero
** calculators*

Page 29 Problem Solving to 999
problem solving showing work
** scratch paper*

Page 30 Code Breaker
vertical forms of subtraction using decimals; creating own code
** calculators, scratch paper*

Page 31 Assessment
** calculators, scratch paper*

Page 32 Subtraction Square Activity

Remember

- ❑ Be aware of using a calculator accurately.
- ❑ Place numerals in the correct columns.
- ❑ Use estimation skills to verify answers.
- ❑ Check work using associated additions for common sense answers, using a calculator.

Additional Activities

- Using computer software catalogues, students spend $1,000, making sure the class needs are met (e.g., boy/girl preferences, learning needs, programs just for fun, etc.). They present their list to the class. Justify choices.
- Create a board game using subtraction skills to 999.
- Observe sporting results and discuss how subtraction skills may be needed.
- Observe subtraction around us—label and display. Report to the class.
- Using supermarket advertisements, research and calculate the balance of your monthly grocery shopping. Give students a set income, (e.g., $800/month).

Answers

Page 26 Subtraction to 999

1. b. 554, c. 811, d. 423, e. 421, f. 260, g. 115, h. 281, i. 600
2. a. 9, 6
   b. 1, 8, 3
   c. 2, 5
   d. 3, 2
3. Across
   1. $30.26
   2. 510
   3. 611.12
   4. 231
   5. 702

   Down
   1. 389
   2. 551
   3. 600
   4. 22
   5. 73

Page 27 Number Smart

1. a. 80, b. 30, c. 550, d. 620, e. 30, f. 70, g. 840, h. 210, i. 90, j. 50, k. 390, l. 740, m. 40, n. 70, o. 100, p. 960, q. 20, r. 80, s. 250, t. 870
2. a. 500, b. 1,000, c. 600, d. 800, e. 500, f. 900, g. 200, h. 600, i. 200, j. 900
3. b. 210, c. 200, d. 180, e. 520, f. 120, g. 770, h. 350, i. 880, j. 70
4. a. 45, 527, 53, 564, 427, 483, 176, 290
   b. 434, 524, 141, 612, 200, 392, 763, 53

Page 28 Trade Away

1. a. 700, b. 300, c. 300, d. 700, e. 200, f. 115, g. 361, h. 508, i. 124, j. 80
2. a. $654, b. $454, c. $277, d. $322, e. $97, f. $1,876, g. $3,286, h. $892, i. $177, j. $2,325
3. a. 20.92, b. 241.56, c. 99.5, d. 522.17, e. 626.26, f. 15.68
4. a. 33.3 m, b. 41.625 L, c. 38.227 kg, d. 32

Page 29 Problem Solving to 999

1. $153.50
2. $19
3. 57% problem solving, 98% subtraction algorithms, 68% calculator usage
4. 586 miles
5. $509.25
6. Check individual work.

Page 30 Code Breaker

1. a. 71.95, b. 800, c. 57.78, d. 60.23, e. 945.46, f. 7.08, g. 291.22, h. 102.18, i. 0.62, j. 8.05, k. 1.55, l. 425.05, m. 270, n. 1.33
   Suzie says she's a super subtractor.
2. Check individual work.

Page 31 Assessment

1. a. 200
   b. 60
   c. 40
   d. 360
   e. 620
   f. 110
2. a. 6, 1
   b. 3, 2, 1
   c. 3, 6, 8
   d. 9, 1
3. a. 581, 921, 24, 199, 710, 321, 437, 353
   b. 615, 453, 279, 330, 25, 565, 127, 14
4. a. 666.65
   b. 77.16
   c. 239.11
   d. 708.4
   e. 5.27
   Ranking d, a, c, b, e
5. Check individual work.

Page 32 Subtraction Square Activity

Check individual work.

Name **Date**

Subtraction and addition are reciprocal operations, e.g., $5 - 3 = 2$ and $2 + 3 = 5$.

1. Subtract, then check with addition.

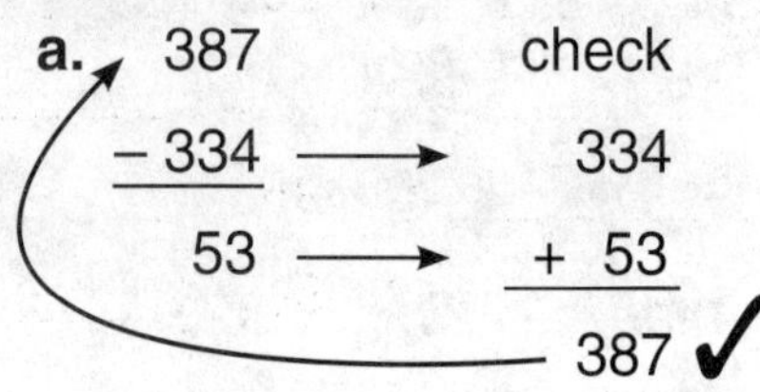

a. 387 − 334 → 53; check: 334 + 53 = 387 ✓

b. 579 − 25 → ___; check: 25 + ___

c. 851 − 40 → ___; check: 40 + ___

d. 436 − 13 → ___; check: 13 + ___

e. 621 − 200 → ___; check: 200 + ___

f. 783 − 523 → ___; check: 523 + ___

g. 246 − 131 → ___; check: 131 + ___

h. 486 − 205 → ___; check: 205 + ___

i. 922 − 322 → ___; check: 322 + ___

2. Complete the algorithms. Use addition to help solve the subtraction sentences.

a. □ 9 9 − 2 □ 3 = 7 3 6

b. □ 7 □ − 4 1 = 1 □ 7

c. 5 8 6 − □ □ = 5 6 1

d. □ 9 3 − 3 4 □ = 5 1

3. Complete the crossnumber puzzle.

Across

1. \$76.66 – \$46.40
2. 350 stamps lost from a collection of 860
3. 618.25 minus 7.13
4. Difference between 266 and 497
5. 759 – 57

Down

1. 100 less than 489
2. 787 subtract 236
3. 711 – 111
4. 369 take away 347
5. \$178 – \$105

Name **Date**

1. Round to the nearest ten.

a. 76	________	**b.** 28	________	**c.** 546	________	**d.** 616	________
e. 32	________	**f.** 73	________	**g.** 837	________	**h.** 208	________
i. 91	________	**j.** 52	________	**k.** 392	________	**l.** 743	________
m. 44	________	**n.** 67	________	**o.** 101	________	**p.** 955	________
q. 15	________	**r.** 79	________	**s.** 254	________	**t.** 867	________

2. Round to the nearest hundred.

a. 494 _____	**b.** 961 _____	**c.** 635 _____	**d.** 780 _____	**e.** 534 _____
f. 853 _____	**g.** 208 _____	**h.** 649 _____	**i.** 192 _____	**j.** 876 _____

3. Estimate by rounding to the nearest ten before subtracting.

a. 293 – 101 =	290 – 100 = 190	**b.** 559 – 347 =	________________
c. 452 – 254 =	________________	**d.** 802 – 616 =	________________
e. 546 – 28 =	________________	**f.** 208 – 91 =	________________
g. 786 – 15 =	________________	**h.** 392 – 44 =	________________
i. 956 – 76 =	________________	**j.** 101 – 32 =	________________

4. Complete the number wheels. Check answers using a calculator.

a.

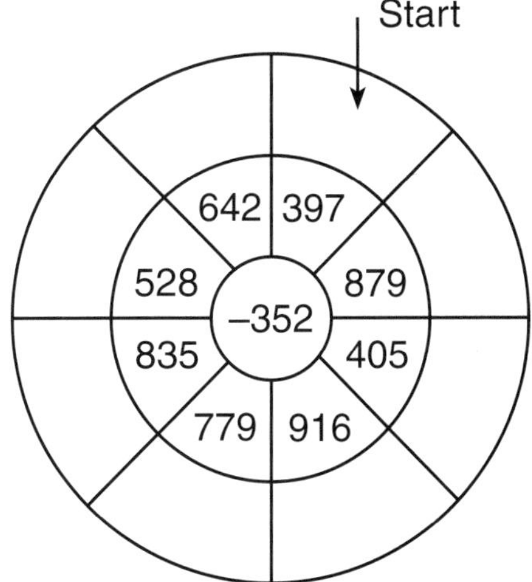

b.

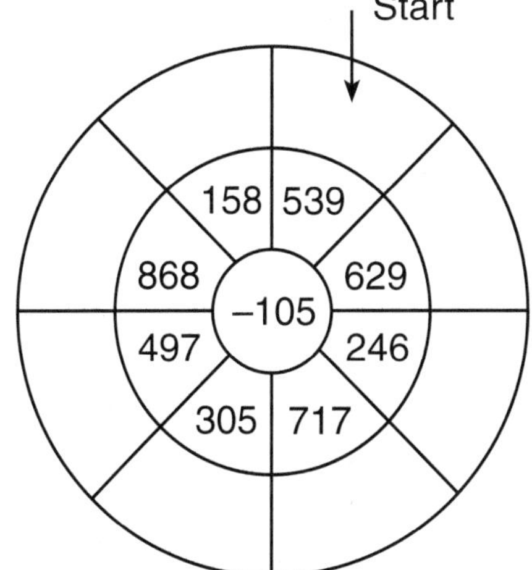

Name **Date**

1. **a.** 800 − 100 **b.** 900 − 600 **c.** 700 − 400 **d.** 900 − 200 **e.** 500 − 300

f. 315 − 200 **g.** 961 − 600 **h.** 708 − 200 **i.** 524 − 400 **j.** 180 − 100

2. **a.** $1,000 − $346 **b.** $700 − $246 **c.** $400 − $123 **d.** $500 − $178 **e.** $2,000 − $1,903

f. $5,000 − $3,124 **g.** $6,000 − $2,714 **h.** $2,700 − $1,808 **i.** $200 − $23 **j.** $3,000 − $675

3. Set these out vertically in the space below before subtracting.

a. 27.16 – 6.24 **b.** 693.22 – 451.66 **c.** 158.34 – 58.84

d. 571.3 – 49.13 **e.** 873.06 – 246.8 **f.** 93.62 – 77.94

a. **b.** **c.**

d. **e.** **f.**

4. Use a calculator to solve these problems. Watch out for the units!

a. 3.56 m – 23 cm = ____________________

b. 42.3 L – 675 mL = ____________________

c. 39.48 kg – 1,253 g = ____________________

d. 12 dozen – 112 = ____________________

Name	Date

	Show Work
1. Sunny Elementary School held a Hot Dog Fundraising Day for the 6th-grade camp trip. Hot dogs were $1.50 each. What amount was raised if 167 students ordered hot dogs? Remember that $97.00 was spent on buying the buns, hot dogs, and condiments.	
2. Jason, his parents, and his younger brother Thomas went to the movies last Tuesday night. Calculate the change from $50 if adult tickets cost $9.50 and children's tickets cost $6.00.	
3. On a math test, Sam scored 70% for problem solving, 86% for subtraction algorithms, and 90% for calculator usage. James had 22% fewer marks for calculator usage, 12% more marks for subtraction algorithms, and 13% less for problem solving. What scores did James receive? ____ problem solving ____ algorithms ____ calculator	
4. Kathy had to travel 802 miles to her grandma's house from her hometown of Arkeun. The odometer showed that she had already driven 216 miles. How much further did she need to travel?	
5. Morgan spent a quarter of her savings to purchase a CD player. She began with $679. How much does she have left now?	

6. On the back of this sheet or on scratch paper, write three word problems and swap them with a friend to solve.

Name **Date**

1. Work these out to crack the code.

a. 73.19 − 1.24 = ___ A	**b.** 856.99 − 56.99 = ___ Y	**c.** 67.12 − 9.34 = ___ H	**d.** 80.28 − 20.05 = ___ I
e. 987.69 − 42.23 = ___ B	**f.** 16.13 − 9.05 = ___ P	**g.** 345.89 − 54.67 = ___ U	**h.** 106.20 − 4.02 = ___ Z
i. 68.24 − 67.62 = ___ R	**j.** 9.3 − 1.25 = ___ E	**k.** 2.75 − 1.2 = ___ T	**l.** 496.2 − 71.15 = ___ C
m. 299.99 − 29.99 = ___ O	**n.** 1.39 − 0.06 = ___ S		

1.33	291.22	102.18	60.23	8.05

1.33	71.95	800	1.33

1.33	57.78	8.05	1.33

71.95

1.33	291.22	7.08	8.05	0.62

1.33	291.22	945.46	1.55	0.62	71.95	425.05	1.55	270	0.62

2. Write your own code using subtraction algorithms on scratch paper and swap with a classmate to reveal a subtraction message. Use a calculator to help you.

Name **Date**

1. Estimate to the nearest ten before subtracting.

a. 234 – 29 = __________ **b.** 103 – 40 = __________ **c.** 941 – 899 = __________

d. 987 – 625 = __________ **e.** 676 – 59 = __________ **f.** 510 – 403 = __________

2. Complete the algorithms. Use addition to help solve the subtraction sentences.

a.

	1	☐	5
–		4	4
	1	2	☐

b.

	9	☐	4	0
–		6	☐	☐
	8	7	1	9

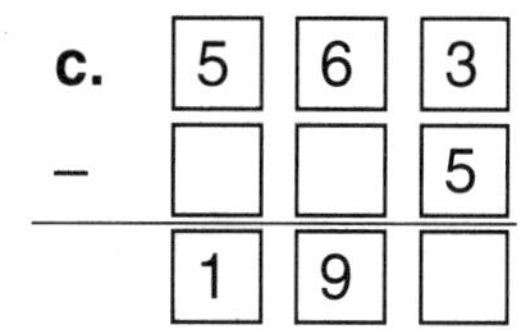

c.

	5	6	3
–	☐	☐	5
	1	9	☐

d.

	7	1	☐
–	2	☐	6
	5	0	3

3. Calculate mentally.

a.

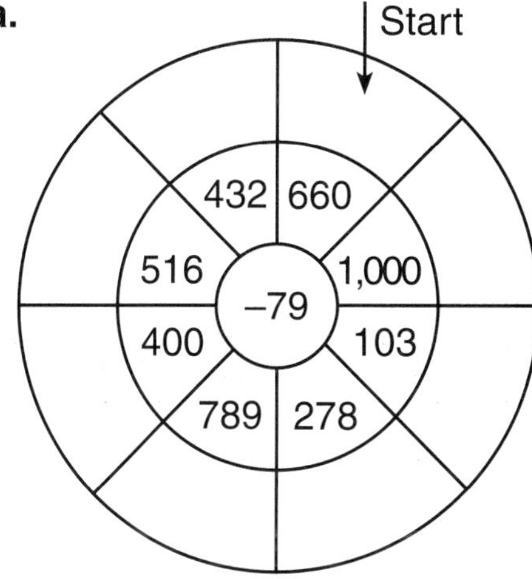

b.

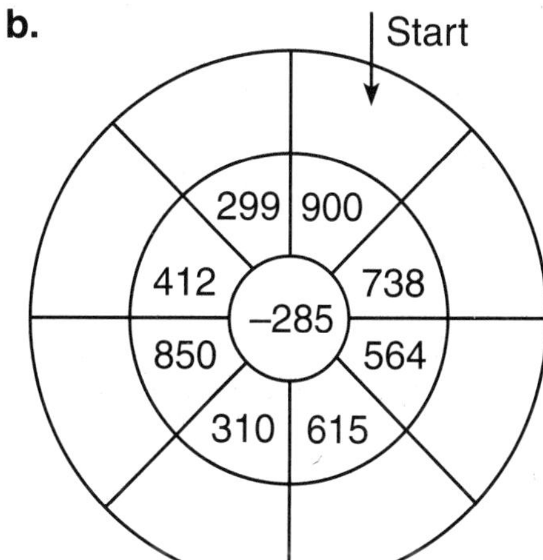

4. Solve these algorithms using a calculator, then rank your answers from largest to smallest.

a. 694.3 – 27.65 = ____________________ Rank: __________

b. 145.23 – 68.07 = ____________________ __________

c. 311.11 – 72.00 = ____________________ __________

d. 727 – 18.6 = ____________________ __________

e. 9.3 – 4.03 = ____________________ __________

5. Write story problems that match the following number sentences on the back of this sheet or on another piece of paper.

a. 565 – 95 = 470 **b.** 200 – 30 = 170

c. 950 – 350 = 600 **d.** 499 – 49 = 450

Name **Date**

This is a magic subtraction square.

- When the middle number of each row, column, or diagonal is subtracted from the sum of the other two numbers, the answer is always the same.
- For this square it is 5.

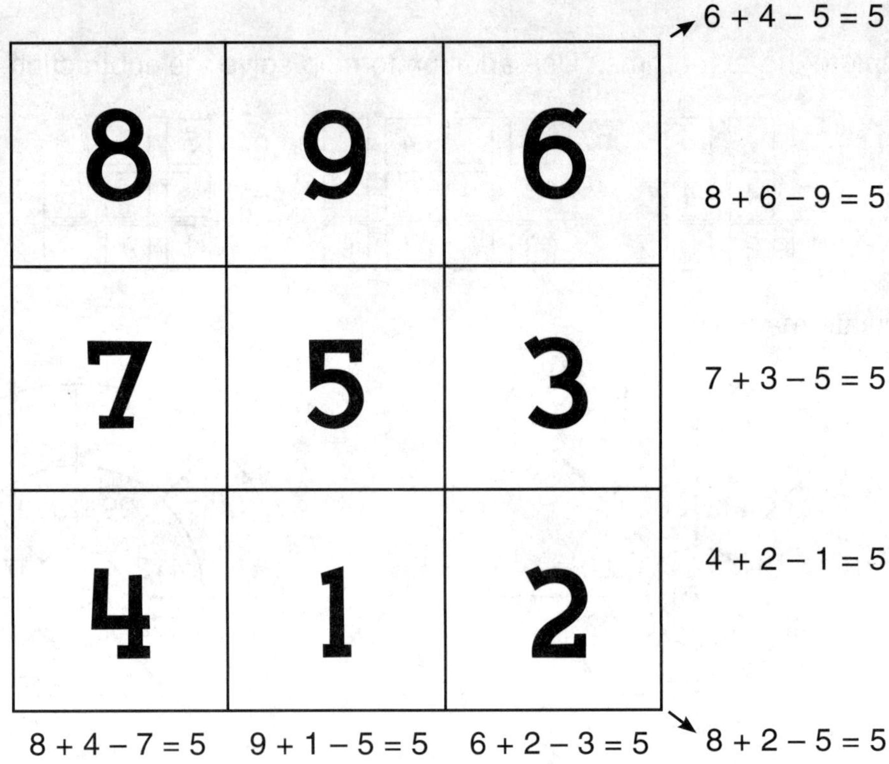

Now it's your turn.

Devise your own magic subtraction square using numbers of your choice.

When you have finished, write in one row only and give it, with a list of the other numbers, to a friend to solve.

WORKING WITH SUBTRACTION

Unit 2

Regrouping
Codes
Subtraction to 9,999
Problem solving

Objectives

- *use a calculator efficiently for operating on numbers*
- *use subtraction skills involving whole numbers to solve problems*
- *select appropriate strategies to approximate and calculate solutions to problems involving whole numbers, money, and decimal fractions*
- *select appropriate technology and use it to help carry out investigations*
- *construct and verify number sentences involving subtraction*
- *identify, continue, and invent whole and fractional number patterns, linked by subtraction*
- *appreciate the impact of mathematical information on daily life*

Language

subtract, regroup, difference, calculate

Materials/Resources

calculators, lined paper

Contents of Student Pages

* *Materials needed for each reproducible student page*

Remember

Encourage students to:

- ❑ *record decimal points directly beneath each other.*
- ❑ *check their own answers by estimating, using common sense, using a calculator, and developing their own examples.*

Additional Activities

- ❑ *Find the population of 10 main cities in one continent of your choice. Graph and compare. Present your graph to the class.*
- ❑ *Research the circulation numbers of a variety of newspapers/magazines. Write comparative statements.*
- ❑ *Create a board game using subtraction skills to 9,999 and beyond.*
- ❑ *Calculate the difference between net and gross mass. Discuss this from the consumers' point of view.*
- ❑ *Research and compare the following, showing how subtraction can be used in everyday situations:*
  - *areas of national parks*
  - *areas of different states*
  - *astronomical distances*
  - *famous people in history, year of birth/death and years lived*
  - *time zones*

Answers

Page 35 Regrouping

1. a. 908 e. 1,760
   b. 1,427 f. 1,189
   c. 2,635 g. 2,372
   d. 2,241 h. 4,013
2. a. 5,568 e. 1,013
   b. 1,343 f. 445
   c. 3,352 g. 1,857
   d. 1,843 h. 1,943
3. a. $1.25 e. $2,638.90
   b. $7.84 f. $7,952.00
   c. $42.60 g. $2,363.80
   d. $746.20 h. $6,577.98
4. $227,500, $313,500, $268,500, $429,500, $179,500, $412,500, $553,500, $311,500

Page 36 Codes

1. a. = e. > or ≠
   b. 75 f. 977
   c. – g. 144
   d. ≠ or > h. 0
2. a. 7,718 e. 8,078
   b. 4,615 f. 7,714
   c. 3,376 g. 7,738
   d. 619

   Bill sighed gleefully after recognizing that the big blob on the hill was only a shadow of a bell.

Page 37 Subtraction to 9,999

1. a. T f. F
   b. F g. T
   c. F h. F
   d. F i. F
   e. T j. T
2. 243 – 158; 735 – 650; 112 – 27; 1,200 – 1,115; 171 – 86; 640 – 555; 2,184 – 2,099; 260 – 175; 941 – 856; 374 – 289; 986 – 901; 185 – 100; 774 – 689; 6,091 – 6,006; 178 – 93; 418 – 333; 647 – 562
3. b. 9,963 – 7,380 = 2,583 or 9,963 – 2,583 = 7,380
   c. 2,596 – 1,596 = 1,000 or 2,596 – 1,000 = 1,596
   d. 9,842 – 4,921 = 4,921
   e. 8,028 – 6,023 = 2,005 or 8,028 – 2,005 = 6,023

Page 38 Mixed Problems

1. a. ($3,520 + $85) – ($550 + $200 + $1,487 + $880) = $488
   b. $1,000 – ($569 + $129) = $302
   c. $1,200 – $750 = $450
   d. 735 – 315 = 420 pairs
   e. $200 – ($49.95 + $89.95) = $60.10
2. a. 119, 69, 44
   b. 1,130, 1,094, 1,085
   c. 7,324, 6,674, 4,724

Page 39 Sales

1. a. Debbie
   b. Debbie $67,750, Mike $63,500, Thomas $62,500, Brad $48,400
   c. $4,250
   d. $19,350
   e. Brad (under) $27,600, Mike (over) $11,500, Debbie (over) $5,250, Thomas (over) $14,500
   f. Yes, (over) $3,650

Page 40 Assessment

1. a. 4,265
   b. 8,428
   c. 3,839.7
   d. 6,132.28
   e. $723.25
   f. $500.45
   g. 233,355
   h. 370,109
2. a. 266, 444, 799, 814
   b. 54, 857, 254, 741, 1,218
3. a. 1.01
   b. 0.45
   c. 9
   d. 20,000
4. Check individual work.
5. Check individual work.

Name **Date**

1. Subtract.

a. 1,908 − 1,000

b. 3,427 − 2,000

c. 5,635 − 3,000

d. 6,241 − 4,000

e. 2,760 − 1,000

f. 4,189 − 3,000

g. 8,372 − 6,000

h. 9,013 − 5,000

2. Borrow to subtract.

a. 7,000 − 1,432

b. 2,000 − 657

c. 6,000 − 2,648

d. 5,000 − 3,157

e. 8,000 − 6,987

f. 1,000 − 555

g. 4,000 − 2,143

h. 3,000 − 1,057

3. Subtract and borrow where necessary.

a. $12.00 − $10.75

b. $20.00 − $12.16

c. $87.65 − $45.05

d. $1,653.00 − $906.80

e. $2,739.26 − $100.36

f. $8,762.00 − $810.00

g. $2,500.00 − $136.20

h. $9,703.44 − $3,125.46

4. If I have a deposit of $37,500, how much money must I borrow from the bank to purchase the following properties?
Check your answers to see if they make sense!

Suburb	Thousands ($)	Loan Amount
Bexlong	265	
Campdon	351	
Castford	306	
Easterway	467	

Suburb	Thousands ($)	Loan Amount
Creachre	217	
Geenly	450	
Lanove	591	
Pyraville	349	

Name **Date**

1. Choose a card to make each number sentence true.

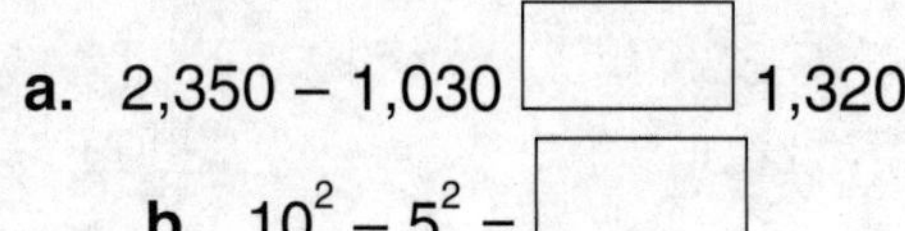
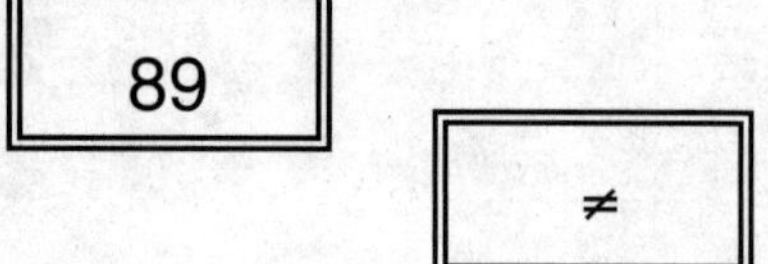

a. 2,350 – 1,030 ______ 1,320

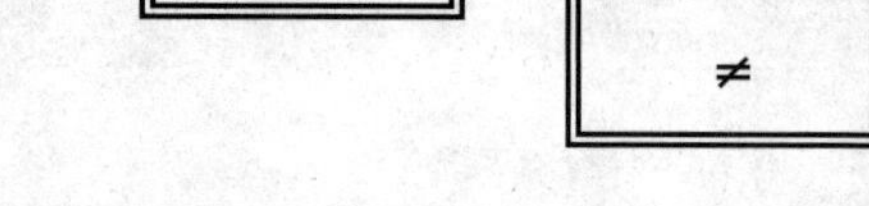

b. $10^2 - 5^2 =$ ______

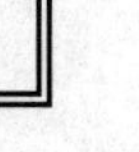

c. \$376.27 + \$876.22 = \$4,525 ______ \$3,272.51

d. 24 + 6 ______ $\sqrt{625} - 1$

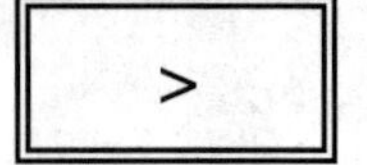

e. 18.8 ______ 29.4 – 10.8

f. ______ – 952 = 25

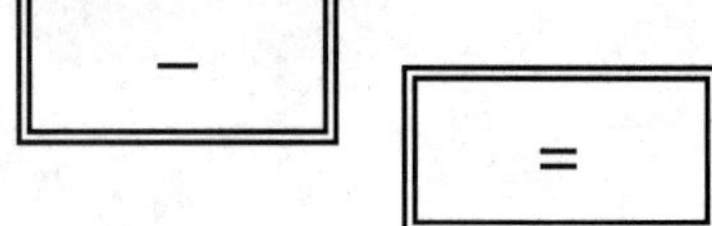

g. $225 - 9^2 =$ ______

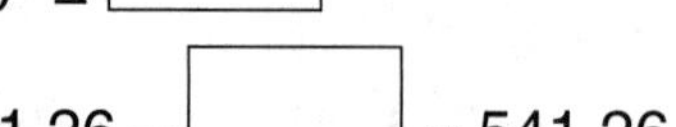

h. 541.26 – ______ = 541.26

Cards: 89, ≠, 977, 144, 75, >, +, –, =, 0

2. Use a calculator to reveal the hidden words in the sentence below. Turn your calculator answers upside down to see the words.

a.	10,032 – 2,314	**b.**	9,837 – 5,222
c.	6,590 – 3,214	**d.**	3,340 – 2,721
e.	9,078 – 1,000	**f.**	8,144 – 430
g.	10,206 – 2,468		

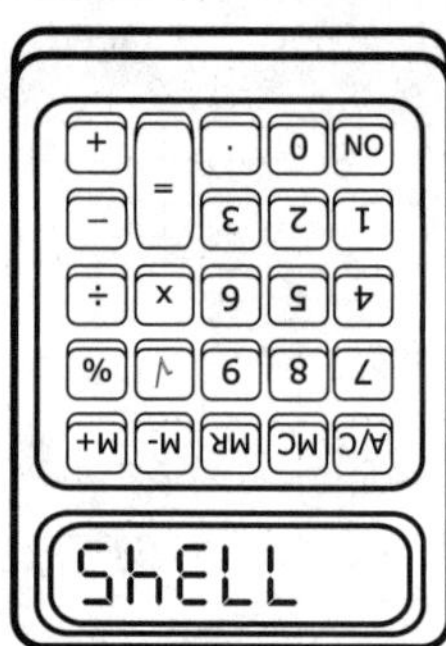

______ (a.) ______ (b.) ed ______ (c.) fully after recognizing that the ______ (d.) ______ (e.) on the ______ (f.) was only a shadow of a ______ (g.).

3. Make your own secret code sentence using subtraction algorithms and swap with a friend.

4. Try creating another code using another form of mathematical technology. No calculator this time!

Name **Date**

1. Circle true (T) or false (F).

a.	69 – 54 < 72 – 41	T/F	**f.**	6,191 – 2,582 = 3,608	T/F
b.	4,865 – 655 = 2,609	T/F	**g.**	258 x 3 = 774 – 0	T/F
c.	1,000 – 555 = 445 + 555	T/F	**h.**	4,770 = 5,770 – 999	T/F
d.	632 + 224 = 865 – 632	T/F	**i.**	921 + 12 + 329 > 1,829	T/F
e.	2,569 = 2,934 – 365	T/F	**j.**	68 = 1,254 – 1,186	T/F

2. An indoor rock climber must scale the wall with caution. Calculate the footholds that equal 85 and climb to the top left of the wall. Shade in his path as you go.

647 – 562	8,321 – 8,321	66 – 29	1,000 – 654	4,331 – 114
418 – 333	178 – 93	6,091 – 6,006	3,952 – 12	688 – 88
9,999 – 99	7,532 – 1,111	774 – 689	960 – 58	51 – 9
374 – 289	986 – 901	185 – 100	123 – 32	256 – 92
941 – 856	300 – 214	779 – 23	864 – 481	621 – 34
260 – 175	2,184 – 2,099	640 – 555	171 – 86	1,200 – 1,115
896 – 63	620 – 554	57 – 882	6,161 – 5,680	112 – 27
993 – 991	444 – 321	321 – 23	543 – 457	735 – 650
833 – 500	407 – 149	425 – 369	197 – 109	243 – 158

3. Write reverse algorithms for these addition sentences.

a. 3,562 + 1,240 = 4,802 4,802 – 3,562 = 1,240 or 4,802 – 1,240 = 3,562

b. 7,380 + 2,583 = 9,963 ____________ or ____________

c. 1,596 + 1,000 = 2,596 ____________ or ____________

d. 4,921 + 4,921 = 9,842 ____________ or ____________

e. 6,023 + 2,005 = 8,028 ____________ or ____________

Name **Date**

1. Write a number sentence to solve each problem.

   a. Nicole has a weekly goal of selling $3,520 worth of shoes at her job. On Monday she sold shoes worth $550, Tuesday $200, Wednesday $1,487, and Friday $880. How much did she sell on Thursday if she exceeded her goal by $85?

   ____

   b. What change would there be from $1,000 if Monique bought a designer dress and a bag for $569 and $129 (tax included)?

   ____

   c. Calculate the difference in price between a diamond ring on sale for $750 and a gold bracelet priced at $1,200.

   ____

   d. Grice Bros. ordered 735 pairs of sunglasses. By the end of the month, they were still awaiting a delivery of 315 pairs. How many sunglasses were delivered earlier in the month?

   ____

   e. James was on strict instructions to keep to a shopping spree limit of $200. How much did he have left to spend after purchasing a shirt for $49.95 and jeans for $89.95 (tax included)?

   ____

2. These steps form a number pattern. Calculate the differences and complete the steps.

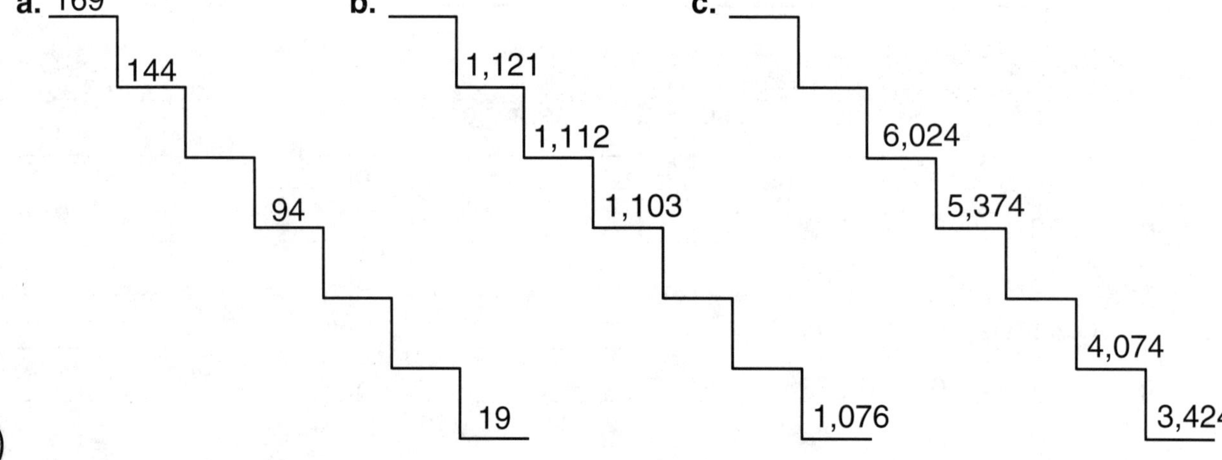

Name	Date

At Sonic Used Car Lot, Thomas, Mike, Debbie, and Brad are the top four salespeople. Look at the table below and answer the questions.

Sonic Used Car Lot Sales Record		
Salesperson	Car	Sale Price
Thomas	1	14,500
	2	21,500
	3	9,000
	4	17,500
Mike	1	26,800
	2	25,200
	3	11,500
Debbie	1	9,000
	2	15,250
	3	7,000
	4	24,500
	5	12,000
Brad	1	14,000
	2	12,000
	3	11,500
	4	10,900

1. **a.** Who sold the most cars this week? ______________________

   b. Order Thomas', Mike's, Debbie's, and Brad's total sales (from highest to lowest).

   ____

   c. What was the difference between Mike's and Debbie's sales? ______________

   d. Calculate the difference between the highest and lowest sales. ______________

   e. Each salesperson sets a target selling goal for the week. Calculate how much they were over or under their target.

Salesperson	Target	Over	Under
Brad	$76,000		
Mike	$52,000		
Debbie	$62,500		
Thomas	$48,000		

   f. If the four salespeople combined their sales, would they have reached their combined target for the week? How much were they over or under their target?

   ____

Name **Date**

1.

a.	7,265 – 3,000	**b.**	13,000 – 4,572	**c.**	4,108 – 268.3	**d.**	6,235.78 – 103.5
e.	\$793.30 – \$70.05	**f.**	\$1,000.90 – \$500.45	**g.**	273,566 – 40,211	**h.**	438,246 – 68,137

2. Complete the cloud patterns.

a.

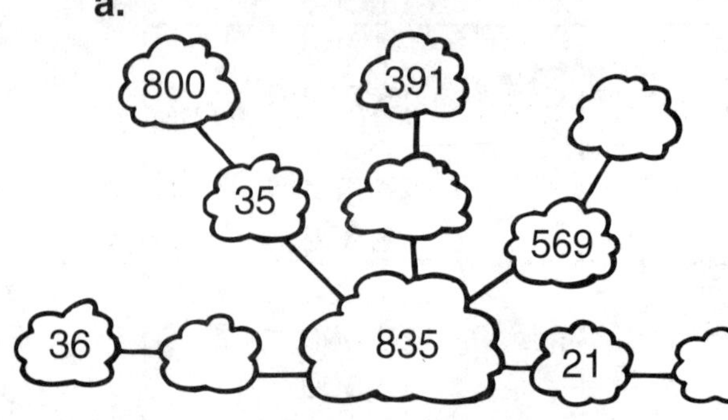

b.

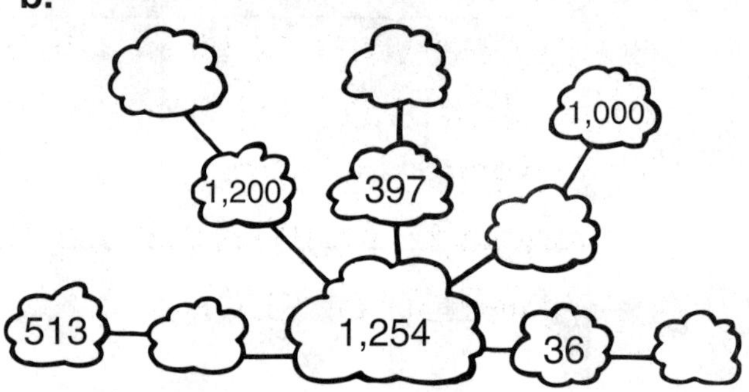

3. Calculate the difference between the two numbers.

   a. 672.00 – 670.99 = ______________________

   b. 36.5 – 36.05 = ______________________

   c. 2,865.1 – 2,856.10 = ______________________

   d. 127,639 – 107,639 = ______________________

4. On an extra sheet of paper, write story problems to match the following number sentences.

   a. \$37,500 – \$8,322 = \$29,178

   b. 500 cars – 200 cars = 300 cars

   c. 87 pounds – 12 pounds = 75 pounds

   d. 416,000 – 18,000 – 23,000 – 12,000 = 363,000

5. Create your own number patterns that involve subtraction.

   a. ____

   b. ____

MULTIPLICATION

Formal setting out of multiplication algorithms is practiced with both vertical and horizontal forms. Algorithms extend to multiplying by three-digit numbers. Estimation is encouraged. Patterns are discovered and extended. Multiplication facts are practiced through the completion of speed grids and finding missing digits.

Problem solving uses multiplication in varied ways. Areas are calculated, money spent, and a school concert is organized. Students are also required to write their own problems to fit given information. Two assessment pages are included.

UNDERSTANDING MULTIPLICATION

Unit 1

Extended form
Contracted form
Patterns
Problem solving
Estimation

Objectives

- *use multiplication involving whole numbers to solve problems*
- *recall and use multiplication facts up to 10 x 10*
- *model and explain patterns related to number facts for multiplication*
- *select appropriate strategies to approximate and calculate solutions to problems involving whole numbers, money, and decimal fractions*
- *appreciate how mathematics is used in a range of aspects of society*
- *use understood written methods to multiply whole numbers to two-digit numbers*
- *use a calculator efficiently for operating on numbers*
- *estimate and/or calculate mentally*

Language

estimate, calculate, units, ones, tens, hundreds, thousands, columns, decimals, decimal point, trade, contracted form, extended form, multiply, multiplication

Materials/Resources

calculators

Contents of Student Pages

* *Materials needed for each reproducible student page*

Page 44 Look and See
patterns; codes
* *calculators*

Page 45 Extended Form
calculating the extended 2-digit x 1-digit algorithm; using columns; horizontal and vertical calculations

Page 46 Contracted Form
calculating the contracted algorithm; estimating
* *calculators*

Page 47 Patterns
multiplying by 1, 10, 100; multiplying decimals; the movement of the decimal point
* *calculators*

Page 48 Concert Problems
estimating; identifying number sentences; calculation of problems

Page 49 Assessment

Remember

Encourage students to:

- ❑ *seek help when unsure.*
- ❑ *explain their understandings.*
- ❑ *use a Times Table Poster if needed.*
- ❑ *use concrete materials if necessary.*

Additional Activities

- *Using supermarket catalogues, explore how multiplication is used.*
- *Play board games using 3 dice and calculators. To move forward multiply the 3 dice and add the number in the total, e.g., 3 x 6 x 4 = 72, 7 + 2 = 9, move forward 9 spaces. If the total was 184, add 1, 8 and 4 to get 13 then add 1 and 3 to move forward 4 spaces.*
- *Students design a simple multiplication board game for lower grades using 2, 3, 4, 5, or 10 times tables.*
- *Daily multiplication grid—students beat their own time if using the same grid outline, developing accuracy and confidence.*

Answers

Page 44 Look and See

1. a. x5: 40, 50, 15, 20, 5, 25, 0, 45, 35, 10, 30
   x10: 80, 100, 30, 40, 10, 50, 0, 90, 70, 20, 60
   b. The x10 row is double the x5 row.
2. a. A = 4 N = 35
   B = 12 O = 10
   C = 40 P = 24
   D = 21 Q = 42
   E = 6 R = 56
   F = 54 S = 9
   G = 36 T = 32
   H = 15 U = 12
   I = 8 V = 63
   J = 27 W = 16
   K = 81 X = 100
   L = 20 Y = 64
   M = 50 Z = 72
   b.–f. Check individual work.

Page 45 Extended Form

1. b. 86 e. 648 h. 627
   c. 88 f. 448
   d. 96 g. 963
2. a. F d. F g. T
   b. T e. T h. F
   c. F f. T i. F
3. a. 260 d. 168 g. 170
   b. 48 e. 189 h. 474
   c. 568 f. 108
4.

21	6	15	27	18	0	12	9	24	30	3
42	12	30	54	36	0	24	18	48	60	6
63	18	45	81	54	0	36	27	72	90	9

Page 46 Contracted Form

1. a. 156 c. 512 e. 108
   b. 234 d. 455 f. 144
2. a. 27 x 6 c. 81 x 8 e. 47 x 7
   b. 32 x 5 d. 15 x 9 f. 13 x 1
3. a. $217.20 d. $333.60
   b. $239.85 e. $105.42
   c. $198.10
4. a. 5,382 d. 3,720
   b. 3,765 e. 364
   c. 2,013
5. a. F c. T e. T
   b. T d. F f. F
6. a. 3 c. 7 e. 3, 1
   b. 6 d. 6, 1

Page 47 Patterns

1. a. 6, 60, 600 c. 22, 220, 2,200
   b. 17, 170, 1,700 d. 39, 390, 3,900
2. a. 300 e. 4,400 i. 670
   b. 950 f. 6,300 j. 7,500
   c. 90 g. 580 k. 180
   d. 820 h. 9,900 l. 260
3. a. 56,000 d. 8,000 g. 9
   b. 38,100 e. 47,000 h. 6,500
   c. 19,800 f. 2,720
4. 3.25 32.5 325
   62.06 620.6 6,206
   97.34 973.4 9,734
   582.11 5,821.1 58,211
5. a. it didn't move
   b. one place to the right
   c. two places to the right
6. 14.49 144.9 1,449 14,490
   30.82 308.2 3,082 30,820
   23.71 237.1 2,371 23,710
   6.432 64.32 643.2 6,432
   10.549 105.49 1,054.9 10,549
   76.58 765.8 7,658 76,580
   322.44 3,224.4 32,244 322,440
   587.90 5,879.0 58,790 587,900

Page 48 Concert Problems

1. 72 signs
2. 170 roses
3. 32 awards
4. 352 students
5. $216
6. 1,650 pages
7. 2,000 shoes

Page 49 Assessment

1. a. 186 c. 468
   b. 128 d. 0
2. a. 170 d. 410
   b. 26,700 e. 82.5
   c. 9,840 f. 376
3. a. correct c. correct
   b. incorrect: 665 d. incorrect: 1,746
4. a. 945 d. 1,516 g. 1,904
   b. 1,968 e. 2,538 h. 5,490
   c. 2,208 f. 1,440
5. a. 152
   b. $119.70
   c. 40.5 feet
6. a. 2 c. 2, 7
   b. 4 d. 9

Name **Date**

1. **a.** Complete the grid.

x	8	10	3	4	1	5	0	9	7	2	6
5											
10											

b. What is the pattern?____

2. **a.** Multiply the calculations from **A** to **Z** then check using a calculator.

A 2 x 2 = ______	**B** 4 x 3 = ______	**C** 8 x 5 = ______	**D** 7 x 3 = ______
E 3 x 2 = ______	**F** 6 x 9 = ______	**G** 9 x 4 = ______	**H** 5 x 3 = ______
I 4 x 2 = ______	**J** 3 x 9 = ______	**K** 9 x 9 = ______	**L** 5 x 4 = ______
M 10 x 5 = ______	**N** 7 x 5 = ______	**O** 5 x 2 = ______	**P** 4 x 6 = ______
Q 6 x 7 = ______	**R** 7 x 8 = ______	**S** 3 x 3 = ______	**T** 4 x 8 = ______
U 6 x 2 = ______	**V** 9 x 7 = ______	**W** 4 x 4 = ______	**X** 10 x 10 = ______
Y 8 x 8 = ______	**Z** 9 x 8 = ______		

b. Calculate the value of your name. ____________________________________

c. Compare this value to three friends' names and rank the four in order from least to greatest.

________________ ________________ ________________ ________________

d. Repeat with your last names.

________________ ________________ ________________ ________________

e. With a partner, try other comparisons such as: teachers' names, street names, football teams, Olympic sports, TV shows, etc.

f. Represent your findings using as many mathematical explanations as you can (e.g. graphs, chance and data, numerical comparisons).

Additional Activities

- *Using supermarket catalogues, explore how multiplication is used.*
- *Play board games using 3 dice and calculators. To move forward multiply the 3 dice and add the number in the total, e.g., 3 x 6 x 4 = 72, 7 + 2 = 9, move forward 9 spaces. If the total was 184, add 1, 8 and 4 to get 13 then add 1 and 3 to move forward 4 spaces.*
- *Students design a simple multiplication board game for lower grades using 2, 3, 4, 5, or 10 times tables.*
- *Daily multiplication grid—students beat their own time if using the same grid outline, developing accuracy and confidence.*

Answers

Page 44 Look and See

1. a. x5: 40, 50, 15, 20, 5, 25, 0, 45, 35, 10, 30
   x10: 80, 100, 30, 40, 10, 50, 0, 90, 70, 20, 60
   b. The x10 row is double the x5 row.
2. a. A = 4 N = 35
   B = 12 O = 10
   C = 40 P = 24
   D = 21 Q = 42
   E = 6 R = 56
   F = 54 S = 9
   G = 36 T = 32
   H = 15 U = 12
   I = 8 V = 63
   J = 27 W = 16
   K = 81 X = 100
   L = 20 Y = 64
   M = 50 Z = 72
   b.–f. Check individual work.

Page 45 Extended Form

1. b. 86 e. 648 h. 627
   c. 88 f. 448
   d. 96 g. 963
2. a. F d. F g. T
   b. T e. T h. F
   c. F f. T i. F
3. a. 260 d. 168 g. 170
   b. 48 e. 189 h. 474
   c. 568 f. 108
4.

21	6	15	27	18	0	12	9	24	30	3
42	12	30	54	36	0	24	18	48	60	6
63	18	45	81	54	0	36	27	72	90	9

Page 46 Contracted Form

1. a. 156 c. 512 e. 108
   b. 234 d. 455 f. 144
2. a. 27 x 6 c. 81 x 8 e. 47 x 7
   b. 32 x 5 d. 15 x 9 f. 13 x 1
3. a. $217.20 d. $333.60
   b. $239.85 e. $105.42
   c. $198.10
4. a. 5,382 d. 3,720
   b. 3,765 e. 364
   c. 2,013
5. a. F c. T e. T
   b. T d. F f. F
6. a. 3 c. 7 e. 3, 1
   b. 6 d. 6, 1

Page 47 Patterns

1. a. 6, 60, 600 c. 22, 220, 2,200
   b. 17, 170, 1,700 d. 39, 390, 3,900
2. a. 300 e. 4,400 i. 670
   b. 950 f. 6,300 j. 7,500
   c. 90 g. 580 k. 180
   d. 820 h. 9,900 l. 260
3. a. 56,000 d. 8,000 g. 9
   b. 38,100 e. 47,000 h. 6,500
   c. 19,800 f. 2,720
4. 3.25 32.5 325
   62.06 620.6 6,206
   97.34 973.4 9,734
   582.11 5,821.1 58,211
5. a. it didn't move
   b. one place to the right
   c. two places to the right
6. 14.49 144.9 1,449 14,490
   30.82 308.2 3,082 30,820
   23.71 237.1 2,371 23,710
   6.432 64.32 643.2 6,432
   10.549 105.49 1,054.9 10,549
   76.58 765.8 7,658 76,580
   322.44 3,224.4 32,244 322,440
   587.90 5,879.0 58,790 587,900

Page 48 Concert Problems

1. 72 signs
2. 170 roses
3. 32 awards
4. 352 students
5. $216
6. 1,650 pages
7. 2,000 shoes

Page 49 Assessment

1. a. 186 c. 468
   b. 128 d. 0
2. a. 170 d. 410
   b. 26,700 e. 82.5
   c. 9,840 f. 376
3. a. correct c. correct
   b. incorrect: 665 d. incorrect: 1,746
4. a. 945 d. 1,516 g. 1,904
   b. 1,968 e. 2,538 h. 5,490
   c. 2,208 f. 1,440
5. a. 152
   b. $119.70
   c. 40.5 feet
6. a. 2 c. 2, 7
   b. 4 d. 9

Name **Date**

1. **a.** Complete the grid.

x	8	10	3	4	1	5	0	9	7	2	6
5											
10											

b. What is the pattern?____________________

____

2. **a.** Multiply the calculations from **A** to **Z** then check using a calculator.

A 2 x 2 = _____ **B** 4 x 3 = _____ **C** 8 x 5 = _____ **D** 7 x 3 = _____

E 3 x 2 = _____ **F** 6 x 9 = _____ **G** 9 x 4 = _____ **H** 5 x 3 = _____

I 4 x 2 = _____ **J** 3 x 9 = _____ **K** 9 x 9 = _____ **L** 5 x 4 = _____

M 10 x 5 = _____ **N** 7 x 5 = _____ **O** 5 x 2 = _____ **P** 4 x 6 = _____

Q 6 x 7 = _____ **R** 7 x 8 = _____ **S** 3 x 3 = _____ **T** 4 x 8 = _____

U 6 x 2 = _____ **V** 9 x 7 = _____ **W** 4 x 4 = _____ **X** 10 x 10 = _____

Y 8 x 8 = _____ **Z** 9 x 8 = _____

b. Calculate the value of your name. ____________________

c. Compare this value to three friends' names and rank the four in order from least to greatest.

__________ __________ __________ __________

d. Repeat with your last names.

__________ __________ __________ __________

e. With a partner, try other comparisons such as: teachers' names, street names, football teams, Olympic sports, TV shows, etc.

f. Represent your findings using as many mathematical explanations as you can (e.g. graphs, chance and data, numerical comparisons).

Name **Date**

1. Multiply, showing every step. Remember to keep numbers in the correct columns.

a.

	T	O	
	9	8	
x		1	
		8	(8 x 1)
+	9	0	(90 x 1)
	9	8	

b.

	T	O
	4	3
x		2
+		

c.

	T	O
	2	2
x		4
+		

d.

	T	O
	3	2
x		3
+		

e.

	H	T	O
	3	2	4
x			2
+			

f.

	H	T	O
	1	1	2
x			4
+			

g.

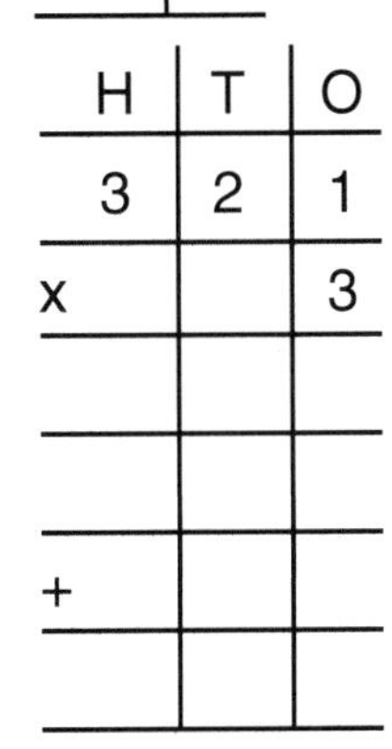

	H	T	O
	3	2	1
x			3
+			

h.

	H	T	O
	6	2	7
x			1
+			

2. True or false?

a. 13 x 3 = 49 ______ **b.** 34 x 2 = 68 ______ **c.** 24 x 4 = 86 ______

d. 123 x 3 = 326 ______ **e.** 349 x 0 = 0 ______ **f.** 863 x 1 = 863 ______

g. 403 x 2 = 806 ______ **h.** 444 x 2 = 666 ______ **i.** 245 x 3 = 645 ______

3. Complete these algorithms using the extended form.

a. 65 x 4 **b.** 16 x 3 **c.** 71 x 8 **d.** 84 x 2 **e.** 27 x 7 **f.** 12 x 9 **g.** 34 x 5 **h.** 79 x 6

4. Complete.

x	7	2	5	9	6	0	4	3	8	10	1
3											
6											
9											

Name **Date**

1. Estimate answers first. Then set each one out vertically to solve.

a. 52 x 3 Est. ____________ **b.** 39 x 6 Est. ____________

c. 64 x 8 Est. ____________ **d.** 91 x 5 Est. ____________

e. 27 x 4 Est. ____________ **f.** 18 x 8 Est. ____________

a. **b.** **c.** **d.** **e.** **f.**

2. Circle the larger answer.

a. 27 x 6 or 18 x 5 **b.** 32 x 5 or 62 x 2

c. 56 x 4 or 81 x 8 **d.** 15 x 9 or 96 x 1

e. 47 x 7 or 101 x 2 **f.** 13 x 1 or 73 x 0

3. Calculate, then check with a calculator.

a.	**b.**	**c.**	**d.**	**e.**
$27.15	$79.95	$39.62	$83.40	$15.06
x 8	x 3	x 5	x 4	x 7

4.

a.	**b.**	**c.**	**d.**	**e.**
897	753	671	465	182
x 6	x 5	x 3	x 8	x 2

5. True or false? Use a calculator.

a. 0.36 x 4 = 144 ________ **b.** 6.27 x 5 = 31.35 ________ **c.** 101.00 x 0 = 0 ________

d. 18.03 x 6 = 108.08 ________ **e.** 20.02 x 7 = 140.14 ________ **f.** 42.15 x 10 = 4.215 ________

6.

a.	**b.**	**c.**	**d.**	**e.**
2 ☐	1 5	3 4	6 9	☐ 5 7
x 2	x ☐	x 5	x 9	x 3
4 6	9 0	1 ☐ 0	☐ 2 ☐	1 0 7 ☐

Name **Date**

1. Use a calculator. Look for the patterns.

a. 6 x 1 = ______ b. 17 x 1 = ______ c. 22 x 1 = ______ d. 39 x 1 = ______
6 x 10 = ______ 17 x 10 = ______ 22 x 10 = ______ 39 x 10 = ______
6 x 100 = ______ 17 x 100 = ______ 22 x 100 = ______ 39 x 100 = ______

2. Now calculate without using a calculator.

a. 3 x 100 = ______ b. 95 x 10 = ______ c. 10 x 9 = ______ d. 10 x 82 = ______
e. 44 x 100 = ______ f. 63 x 100 = ______ g. 58 x 10 = ______ h. 99 x 100 = ______
i. 67 x 10 = ______ j. 75 X 100 = ______ k. 10 x 18 = ______ l. 10 x 26 = ______

3. Now try these.

a. 56 x 1,000 = ______ b. 381 x 100 = ______ c. 198 x 100 = ______ d. 80 x 100 = ______
e. 47 x 1,000 = ______ f. 272 x 10 = ______ g. 9 x 1 = ______ h. 65 x 100 = ______

4. Use a calculator and watch the answers carefully.

	x 1	x 10	x 100
3.25			
62.06			
97.34			
582.11			

5. Where did the decimal point move to when you:

a. multiplied by 1? ______________________
b. multiplied by 10? ______________________
c. Multiplied by 100? ______________________

6. Complete these. No calculators!

	x 1	x 10	x 100	x 1,000
14.49				
30.82				
23.71				
6.432				
10.549				
76.58				
322.44				
587.90				

Name **Date**

Miss Major-Mayhem is organizing the school concert. Help her finalize the arrangements. Estimate first, then calculate to see if your estimate makes sense.

Show Work

1. Three rows of VIP seats are to be set out with 24 seats in each row. How many VIP reserved signs need to be made?

   Estimation ____________ Calculation____________

2. Nathan's mom is making 17 flower arrangements for decoration. Each arrangement consists of 10 roses. How many roses must she order?

   Estimation ____________ Calculation____________

3. The Consistent Effort Awards are being made. Kindergarten, Grade 1, and Grade 2 need four each. Grades 3–6 need five each. How many need to be printed?

   Estimation ____________ Calculation____________

4. Miss Major-Mayhem said that eight rows of 44 seats would be needed for the school students. How many students attend the school?

   Estimation ____________ Calculation____________

5. Hall hiring costs $72 per hour. The concert will take 3 hours. How much will Miss Major-Mayhem write a check for?

   Estimation ____________ Calculation____________

6. Three pages are used per program. How many pages are needed for 550 programs?

   Estimation ____________ Calculation____________

7. With over 1,000 people expected to fill the concert hall, each wearing one pair of shoes, what is the least amount of shoes expected to be in the building?

   Estimation ____________ Calculation____________

Name ____________________ **Date** ____________________

1\. a. 62 x 3

b. 32 x 4

c. 234 x 2

d. 567 x 0

2\. a. 17 x 10 = ______ b. 267 x 100 = ______ c. 984 x 10 = ______

d. 4.1 x 100 = ______ e. 8.25 x 10 = ______ f. 3.76 x 100 = ______

3\. Find and correct any errors by rewriting each one below.

a. 72 x 6 = 432

b. 95 x 7 = 102

c. 916 x 4 = 3,664

d. 582 x 3 = 245

a. b. c. d.

4\. Multiply and show your work. Rewrite a–d vertically.

a. 189 x 5
b. 246 x 8
c. 552 x 4
d. 379 x 4

a. b. c. d.

e. 423 x 6

f. 720 x 2

g. 272 x 7

h. 610 x 9

5\. a. At a local neighborhood meeting, 4 rows of chairs were set up with 38 in each row. How many people could be seated? ______

b. Nicole sold six pairs of shoes on sale for $19.95 each. What was the total amount of her sales (not including tax)? ______

c. Silvana received an order to cut and sew nine tablecloths 4.5 feet in length. How much material is needed to complete the order? ______

6\. Fill in the missing digits.

a.  x 5 = 3 6 0

b. 3 6 0 x ☐ = 1, 4 4 0

c. 1 5 9 x 8 = 1 ☐ ☐ 2

d. ☐ 4 x 3 = 2 8 2

WORKING WITH MULTIPLICATION

Unit 2

Estimating
Patterns
Algorithms
Problem solving
Calculator
Two x two digits

Objectives

- *recall multiplication facts up to 10 x 10*
- *model and explain patterns related to number facts for multiplication*
- *select appropriate strategies to approximate and calculate solutions to problems*
- *use understood written methods to multiply two-digit numbers by two-digit numbers*
- *use a calculator efficiently for operating on numbers*
- *describe situations which may be represented by given, symbolically expressed, number sentences*
- *construct and verify number sentences involving multiplication*
- *estimate and/or calculate mentally*

Language

estimate, calculate, units, ones, tens, hundreds, thousands, columns, decimals, decimal point, trade, contracted algorithms, extended algorithms, number patterns, multiply, multiplication

Materials/Resources

calculators, colored pencils, scratch paper

Contents of Student Pages

* *Materials needed for each reproducible student page*

Remember

Before starting, ensure that students:

- ❑ *seek help when unsure.*
- ❑ *explain their understandings.*
- ❑ *use a Times Table Poster if needed.*
- ❑ *develop good work habits of motivation, perseverance, and independence.*
- ❑ *work together.*
- ❑ *evaluate their learning.*

Additional Activities

- ❑ *Discuss and construct a display (posters, models, etc.) which reflects the theme: "Multiplication is alive and well in our society."*
- ❑ *Conduct a debate . . . "It is essential to learn your Times Tables."*
- ❑ *Design a simple multiplication board game for lower grades using 2, 3, 4, 5, or 10 times tables.*
- ❑ *Design a multiplication board game for 5th and 6th graders.*
- ❑ *Daily multiplication grid. Students can beat their own time if using the same grid outline, thus developing accuracy and confidence.*

Answers

Page 52 Estimates

1. a. 10 x 20 b. 40 x 20 c. 20 x 30 d. 10 x 20 e. 50 x 40 f. 50 x 50 g. 60 x 30 h. 20 x 70 i. 80 x 20 j. 20 x 30
2. a. 1,476 b. 1,898 c. 2,304 d. 935 e. 2,116 f. 407 g. 420 h. 216 i. 910 j. 289
3. a. 600 b. 3,500 c. 800 d. 1,600 e. 2,400
4. a. 15 x $10 = $150
   b. 30 x $20 = $600
   c. $30 x 40 = $1,200
5. Check individual work.

Page 53 Patterns

1. a. 6, 60, 600, 6,000
   b. 4, 40, 400, 4,000
   c. 7, 70, 700, 7,000
   d. 3, 30, 300, 3,000
   e. An additional zero gets added at each step.
2. a. 8, 80, 800
   b. 72, 720, 7,200
   c. 12, 120, 1,200
   d. 25, 250, 2,500
   e. The number of zeros in the question equals the number of zeros in the answer.
3. a. 80, 800, 8,000
   b. 720, 7,200, 72,000
   c. 120, 1,200, 12,000
   d. 250, 2,500, 25,000
   e. There is an additional zero as you go down the pattern.
4. a. 240 b. 1,800 c. 240 d. 48,000 e. 2,100 f. 2,700 g. 1,000 h. 630 i. 1,000 j. 20,000 k. 2,400 l. 100 m. 240 n. 720 o. 2,000 p. 1,500
5. a. 21.1, 211, 2,110
   b. 211, 21,100, 2,110,000
   c. 325.6, 3,256, 32,560
   d. 3,256, 325,600, 32,560,000
   e. When you multiply by 10, the decimal moves one place to the right. When you multiply by 100, the decimal moves two places to the right.

6. Check individual work.

Page 54 2 x 2-Digit Algorithms

1. b. 90, 900, 990
   c. 355, 2,130, 2,485
   d. 234, 260, 494
   e. 85, 1,360, 1,445
   f. 0, 1,440, 1,440
   g. 68, 1,700, 1,768
   h. 248, 1,860, 2,108
2. a. circle 456, 5,380; 462, 5,390, 5,852
   b. circle 92, 1,342; 492, 820, 1,312
   c. circle 56, 520, 569; 24, 360, 384
   d. circle 50, 0, 500, 500
3. a. 108
   b. 480
   c. $22.50

Page 55 Revision

a. 776 b. 1,664 c. 3,510 d. 1,512 e. 450 f. 5,656 g. 1,722 h. 2,488 i. 4,662 j. 1,700 k. 412 l. 1,984 m. 1,048 n. 0 o. 2,250 p. 1,633 q. 800 r. 1,884 s. 198 t. 1,392 u. 2,376 v. 1,353 w. 6,900 x. 2,016

Thank you all!

Page 56 Assessment

1. a. 288 b. 657 c. 744 d. 4,510 e. 5,117
2. a. 35, 350, 3,500
   b. 600, 560, 2,700
   c. 1,800, 2,000, 8,000
   d. 320, 6,400, 4,800
3. a. 74.7 b. 19,600 c. 4,101 d. $315 e. 211.3 f. 24,000 g. 6,520 h. $7.50 i. 539 j. 3,870 k. 89,100 l. $520.50
4. a. Yes b. No c. No d. Yes e. No f. No g. Yes
5. a. 64, 320, 384
   b. 129, 860, 989
   c. 310, 3,100, 3,410
   d. 584, 1,460, 2,044
   e. 51, 1,530, 1,581
6. Check individual work.

Name **Date**

1. Write an estimated number sentence by rewriting the problems below and rounding to the tens place.

a. 12 x 18 ⟶ ______________ b. 36 x 16 ⟶ ______________

c. 18 x 32 ⟶ ______________ d. 14 x 17 ⟶ ______________

e. 45 x 36 ⟶ ______________ f. 48 x 54 ⟶ ______________

g. 63 x 30 ⟶ ______________ h. 24 x 72 ⟶ ______________

i. 81 x 23 ⟶ ______________ j. 18 x 31 ⟶ ______________

2. Estimate, then circle the best answer. Check your own answers using a calculator.

a. 82 x 18	4,176	1,476	542
b. 26 x 73	819	412	1,898
c. 64 x 36	2,304	627	4,032
d. 17 x 55	4,056	935	35,234
e. 46 x 46	342	2,116	688
f. 11 x 37	407	201	50
g. 28 x 15	1,566	967	420
h. 12 x 18	857	1,899	216
i. 91 x 10	349	910	1,999
j. 17 x 17	289	1,690	27,222

3. Estimate an answer for these algorithms.

a. 31 x 18 Est. ________ **b.** 73 x 48 Est. ________ **c.** 44 x 17 Est. ________ **d.** 80 x 23 Est. ________ **e.** 56 x 39 Est. ________

4. Write an estimated number sentence and answer for these problems.

a. Fifteen bunches of flowers were delivered on Friday at $12 per bunch. How much did they cost? ______________________________

b. Kylie booked in 27 photographic sessions valued at $19.95 each. What is the total value? ______________________________

c. Roy organized a trail bike competition. The entry fee is $26 per member. Thirty-seven members have entered the competition. How much will he collect? ________

5. On a separate sheet of paper, write your own story problems for these estimated number sentences.

a. 20 x 60 **b.** 40 x 10 **c.** 80 x 100 **d.** 50 x 30

Name **Date**

1. **a.** 6 x 1 = ______ 6 x 10 = ______ 6 x 100 = ______ 6 x 1,000 = ______
   b. 4 x 1 = ______ 4 x 10 = ______ 4 x 100 = ______ 4 x 1,000 = ______
   c. 7 x 1 = ______ 7 x 10 = ______ 7 x 100 = ______ 7 x 1,000 = ______
   d. 3 x 1 = ______ 3 x 10 = ______ 3 x 100 = ______ 3 x 1,000 = ______

2. **a.** 2 x 4 = ______ 2 x 40 = ______ 2 x 400 = ______
   b. 8 x 9 = ______ 8 x 90 = ______ 8 x 900 = ______
   c. 6 x 2 = ______ 6 x 20 = ______ 6 x 200 = ______
   d. 5 x 5 = ______ 5 x 50 = ______ 5 x 500 = ______
   e. What pattern can you see? ______

3. **a.** 20 x 4 = ______ 20 x 40 = ______ 20 x 400 = ______
   b. 80 x 9 = ______ 80 x 90 = ______ 80 x 900 = ______
   c. 60 x 2 = ______ 60 x 20 = ______ 60 x 200 = ______
   d. 50 x 5 = ______ 50 x 50 = ______ 50 x 500 = ______
   e. Explain the pattern involving zeros. ______

4. **a.** 6 x 40 = ______ **b.** 2 x 900 = ______ **c.** 4 x 60 = ______ **d.** 60 x 800 = ______
   e. 70 x 30 = ______ **f.** 90 x 30 = ______ **g.** 200 x 5 = ______ **h.** 9 x 70 = ______
   i. 10 x 100 = ______ **j.** 50 x 400 = ______ **k.** 80 x 30 = ______ **l.** 10 x 10 = ______
   m. 12 x 20 = ______ **n.** 18 x 40 = ______ **o.** 25 x 80 = ______ **p.** 50 x 30 = ______

5. Follow the pattern. You may check your answers with a calculator.
   a. 2.11 x 10 → ______ x 10 → ______ x 10 → ______
   b. 2.11 x 100 → ______ x 100 → ______ x 100 → ______
   c. 32.56 x 10 → ______ x 10 → ______ x 10 → ______
   d. 32.56 x 100 → ______ x 100 → ______ x 100 → ______
   e. There is a pattern involving the decimal point. Explain it.

______

6. Make up your own pattern of number sentences and swap them with a friend.
   a. ______ ______ ______
   b. ______ ______ ______
   c. ______ ______ ______

Name **Date**

1. a.
```
    32
x   13
    96 (32 x 3)
+  320 (32 x 10)
   416
```

b.
```
    45
x   22

+
```

c.
```
    71
x   35

+
```

d.
```
    26
x   19

+
```

e.
```
    17
x   85

+
```

f.
```
    16
x   90

+
```

g.
```
    34
x   52

+
```

h.
```
    62
x   34

+
```

2. Circle the errors and recalculate correctly below.

a.
```
     77
x    76
    456
+ 5,380
  5,852
```

b.
```
     82
x    16
    492
+    92
  1,342
```

c.
```
     12
x    32
     56
+   520
    569
```

d.
```
     50
x    10
     50
+   500
    500
```

a. b. c. d.

3. Solve these problems.

a. Betty bought 18 dozen eggs for the Egg and Spoon Race. How many more need to be colored if 108 are already colored? __________

b. A donation of 1,200 candies were used to make up 280 grab bags. Laurie decided to have 6 candies per bag. Calculate how many more candies are needed to make up the bags. __________

c. William took 750 pictures to sell at the fair, but only 375 were sold at 50 cents each. If it cost him 22 cents each to print each picture, how much profit was made?

____

Name **Date**

Help Melissa switch on the monitors that work. Color in the monitors where the total is between 1,500 and 2,500. What was Melissa's message? Use the back of this page or a separate sheet of paper to show your work.

a. 

b.
832
x 2
T

c.
54
x 65
E

d. 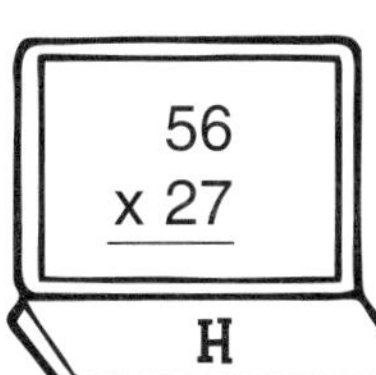

e. 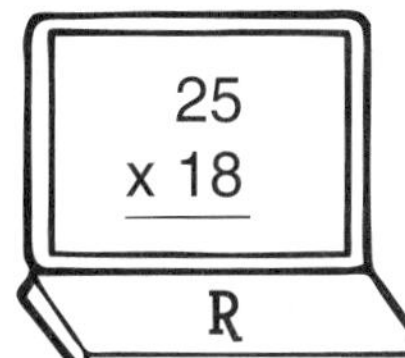

f. 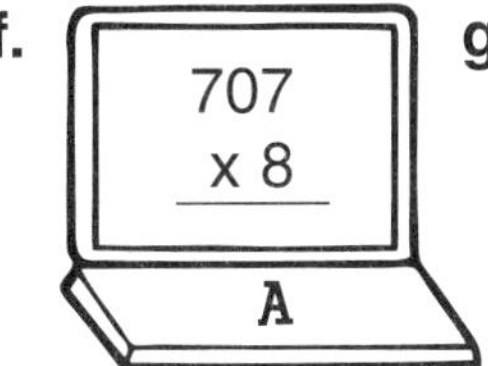

g.
246
x 7
A

h. 

i. 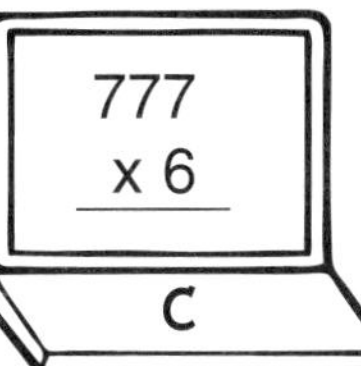

j. 

k. 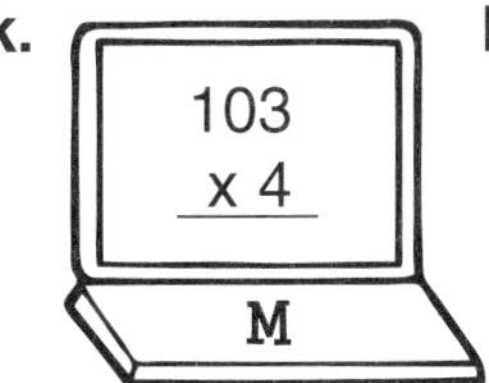

l.
32
x 62
Y

m. 

n. 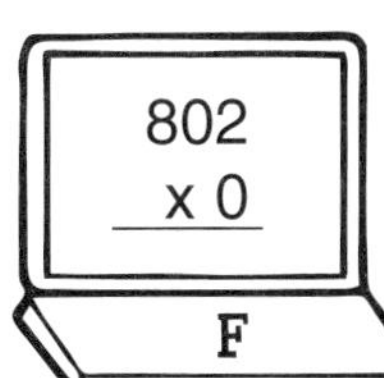

o. 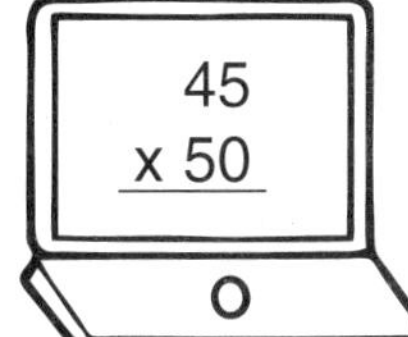

p. 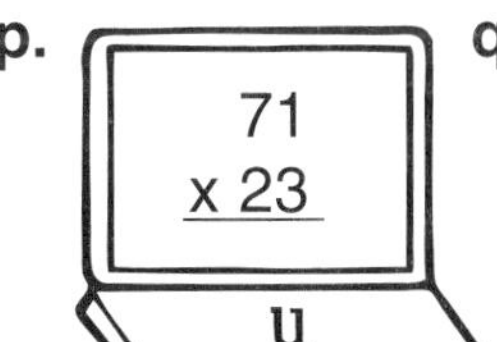

q. 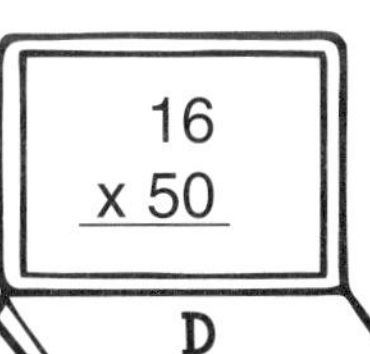

r.
314
x 6
A

s. 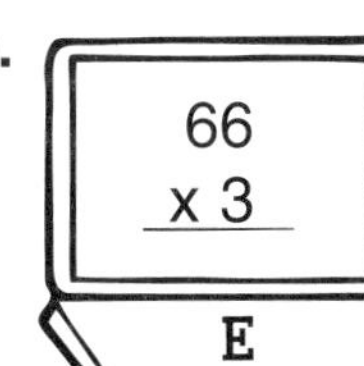

t. 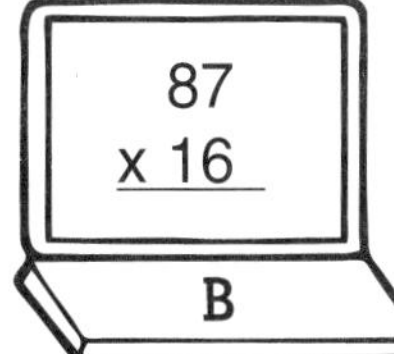

u. 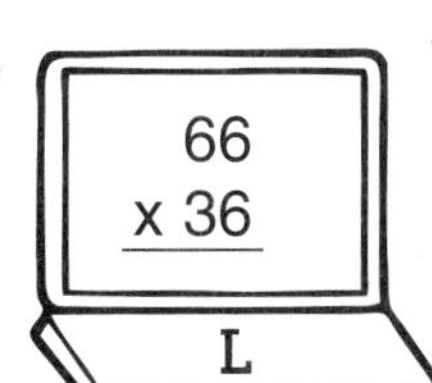

v. 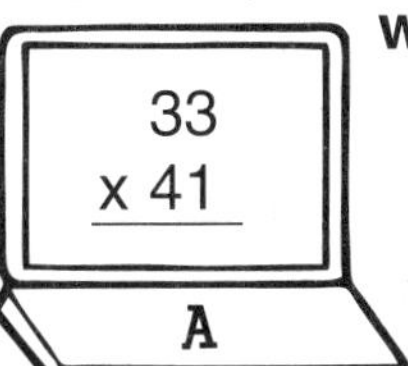

w. 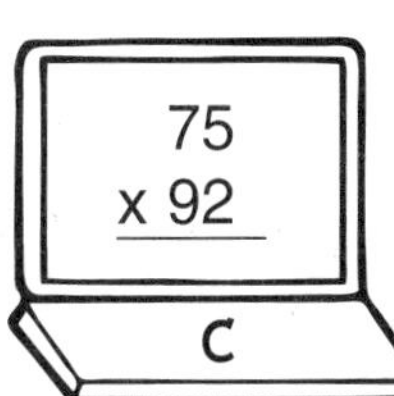

x. 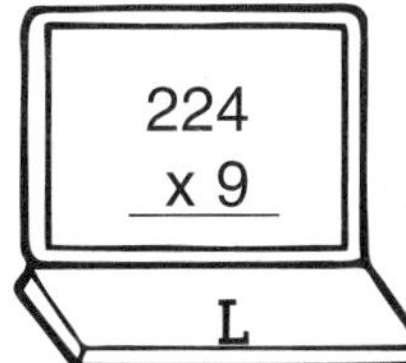

____ ____ ____ ____ ____   ____ ____ ____   ____ ____ ____!

1664 1512 1722 2488 1700   1984 2250 1633   1884 2376 2016

Name **Date**

1. **a.** 48 x 6 **b.** 73 x 9 **c.** 248 x 3 **d.** 902 x 5 **e.** 731 x 7

2. **a.** 5 x 7 = ______ 5 x 70 = ______ 5 x 700 = ______
b. 30 x 20 = ______ 8 x 70 = ______ 3 x 900 = ______
c. 90 x 20 = ______ 40 x 50 = ______ 400 x 20 = ______
d. 8 x 40 = ______ 80 x 80 = ______ 800 x 6 = ______

3. **a.** 7.47 x 10 = ______ **b.** 196 x 100 = ______
c. 41.01 x 100 = ______ **d.** $3.15 x 100= ______
e. 21.13 x 10 = ______ **f.** 24 x 1,000 = ______
g. 65.2 x 100 = ______ **h.** $0.75 x 10 = ______
i. 53.9 x 10 = ______ **j.** 3.87 x 1,000 = ______
k. 89.1 x 1,000 = ______ **l.** $52.05 x 10 = ______

4. Estimate to see if the answer is correct. Circle your answer. Write your estimate.

		Estimate
a. 85 x 95 = 8,075	Yes No	______
b. 21 x 68 = 7,033	Yes No	______
c. 58 x 30 = 105	Yes No	______
d. 94 x 67 = 6,298	Yes No	______
e. 11 x 27 = 721	Yes No	______
f. 95 x 36 = 1,036	Yes No	______
g. 10 x 48 = 480	Yes No	______

5. Show your work.

a. 32 x 12 + **b.** 43 x 23 + **c.** 62 x 55 + **d.** 73 x 28 + **e.** 51 x 31 +

6. Write a story problem for the number sentence 26 x 15 = 390.

______

______

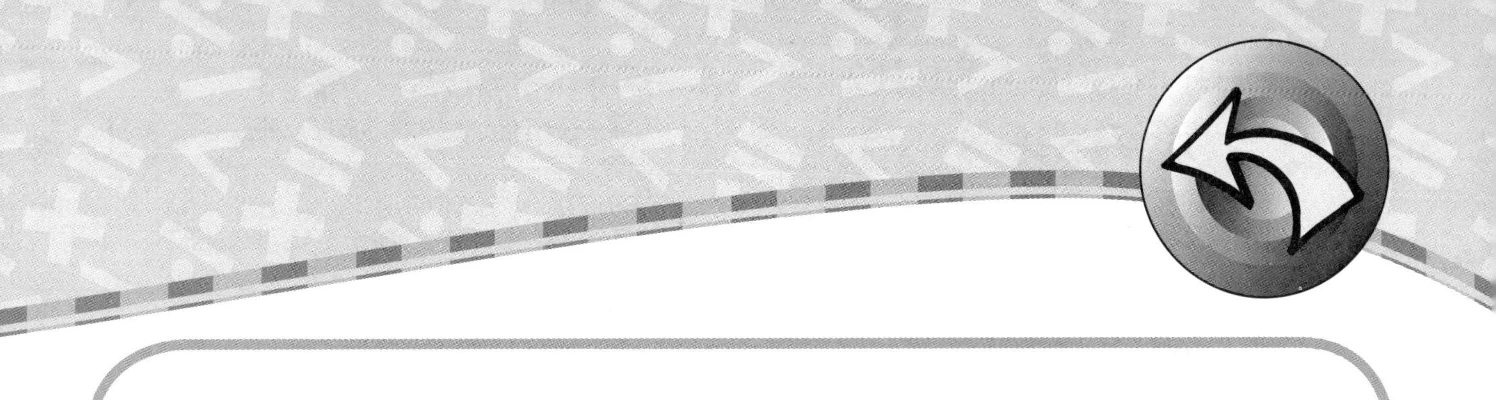

DIVISION

The skills of grouping, sharing, dividing, and calculating are used in these division units. Formal algorithms are worked, with and without remainders, and extend to two-digit divisors.

Problems are solved using division with measurements and quantities, and students are encouraged to write their own problems involving division. Estimation is encouraged in all phases, and calculators are used to check answers. Fun pages including code breaking, crossnumber puzzle, patterning, and choosing the correct path all help reinforce division number facts.

There are two assessment pages included, and the activity page involves using rules of divisibility to trace a path for Fred, the frog!

UNDERSTANDING DIVISION

Unit 1

Factors
Remainders
Patterns
Puzzles and codes
Problem solving

Objectives

- *use division involving whole numbers to solve problems*
- *recall and use division facts up to 10 x 10*
- *use understood written methods to divide whole numbers by single-digit numbers*
- *use a calculator efficiently for operating on numbers*
- *estimate and calculate mentally*
- *identify and continue number patterns*

Language

share, group, divide, divided by, division, remainder, factors, quotient, average, single-digit division

Materials/Resources

calculators, scratch paper

Contents of Student Pages

* *Materials needed for each reproducible student page*

Page 60 Basic Division
sequential order of division facts; factors; averages
* *calculators, scratch paper*

Page 61 Remainders
setting out division equations; sequential order of division facts; calculating measurements; writing own story problems
* *scratch paper*

Page 62 Crossnumber
division crossnumber puzzle

Page 63 Wheel of Fun
cracking a code to reveal a message

Page 64 Patterns
dividing by 1, 10, and 100; dividing by zero; relationship between division and multiplication; division paths
* *calculators*

Page 65 Assessment

Page 66 Division Activity
finding a path using divisibility rules
* *calculators*

Remember

Encourage students to:

- ❑ *seek help when unsure.*
- ❑ *explain their understandings.*
- ❑ *use a Times Table Poster if needed.*
- ❑ *use concrete materials if necessary.*
- ❑ *use calculators to check their work.*

Additional Activities

- *Use supermarket catalogues and explore how division is used.*
- *Daily division grid. Students can beat their own time if they use the same grid outline, thus developing accuracy and confidence.*
- *Design a division board game.*
- *Brainstorm and display things that can be shared.*

Answers

Page 60 Basic Division

1. a. 123 b. 123 c. 121 d. 111 e. 110 f. 100 g. 110 h. 101 i. 101 j. 101 k. 131 l. 162 m. 141 n. 140 o. 172 p. Check individual work.
2. a. 1, 17
   b. 1, 3, 9, 27, 81
   c. 1, 5, 25, 125
   d. 1, 2, 4, 7, 8, 14, 16, 28, 32, 56, 112, 224
   e. 1, 2, 4, 5, 10, 20, 25, 50, 100, 125, 250, 500
3. a. 20 b. 36 c. 75 d. 362 e. 456
4. a. 60.25 b. 81.75 c. 61.5 d. 70

Page 61 Remainders

1. a. $3.03 b. $22.68 c. $147.20 d. $65.20 e. $14.60
2. a. 102 r2 b. 301 r2 c. 102 r3 d. 111 r1 e. 69 r4 f. 125 r2 g. 223 r5 h. 96 r1 i. 123 r3 j. 101 r8 k. 136 l. 52 r5 m. 56 r3 n. 98 r2 o. 84 r3
3. a. 600 mL b. 825 g c. 7.5 m d. 9 days e. 12
4. a. 1 b. 5 c. 3 d. 4
5. a. T b. F c. F d. T e. T f. F g. F h. T i. F
6. Check individual work.

Page 62 Crossnumber

Across

1. 10
2. 36
3. 80
4. 61
5. 15
6. 564
7. 500
8. 48
9. 40
10. 81
11. 217
12. 19
13. 2,500
15. 9,000
16. 192
17. 20
18. 160
19. 12
21. 5,921
23. 622
25. 27

Down

1. 10
2. 344
3. $8.10
4. 67
5. $12.30
6. 50
7. 561
8. 472
9. 400
10. 89
11. 212
12. 10
14. 570
15. 901
16. 10
17. 211
18. 122
20. 250
22. 200
23. 60
24. 223

Page 63 Wheel of Fun

T. 44
N. 31
M. 26
S. 82
F. 58
H. 73
A. 92
C. 17
L. 42
Y. 36
P. 95
Q. 18
R. 62
U. 23
I. 15
E. 59

Ten Minutes Free Play!

Page 64 Patterns

1.

810	560	777	403	200	640	392	930
81	56	77.7	40.3	20	64	39.2	93
8.1	5.6	7.77	4.03	2	6.4	3.92	9.3

2. It has no meaning. You can't divide something by nothing.
3. a. 141 x 4 = 564
   b. 111 x 9 = 999
   c. 201; 201 x 8 = 1,608
   d. 81; 81 x 5 = 405
4. a. 2,520, 504, 72, 18
   b. 5,520, 2,760, 115, 23
   c. 2,268, 252, 42, 14
   d. 3,744, 936, 156, 26
   Path B

Page 65 Assessment

1. a. 231 b. 81 c. 451 d. 28 e. 84
2. a. 4 b. 9 c. 31 d. 1 e. Check individual work.
3. a. 1,608 b. 1,902 c. 156 d. 520 e. 983
   order: b, a, e, d, c
4. a. 2 b. 0 c. 2, 4 d. 6 e. 7, 8
5. a. F b. T c. F d. T e. T f. F g. T h. F
6. a. 768 ÷ (8 x 6) = 16 trips
   b. 768 ÷ 4 = 192 cars

Page 66 Division Activity

1. The path is 121; 1,792; 963; 623; 979; 737; 2,429; 2,002; 801; 189; 1,177; 3,384; 522; 1,099; 946; 605; 1,404; 585; 602; 1,736; 5,643; 266; 1,548; 6,820; 1,029; 704

Name **Date**

1. **a.** $3\overline{)369}$ **b.** $2\overline{)246}$ **c.** $4\overline{)484}$ **d.** $5\overline{)555}$ **e.** $7\overline{)770}$

   f. $5\overline{)500}$ **g.** $9\overline{)990}$ **h.** $6\overline{)606}$ **i.** $7\overline{)707}$ **j.** $8\overline{)808}$

   k. $5\overline{)655}$ **l.** $2\overline{)324}$ **m.** $4\overline{)564}$ **n.** $6\overline{)840}$ **o.** $3\overline{)516}$

   p. Make up five division problems. Rewrite them on a sheet of scratch paper for a friend to solve.

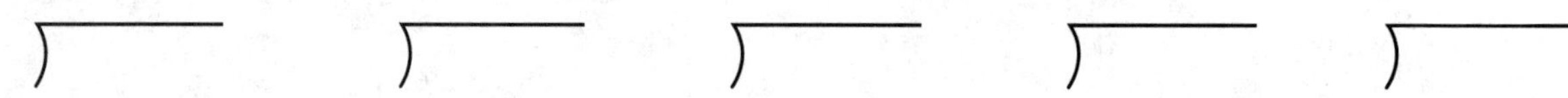

2. Write the factors of these numbers.

   a. 17 ______________________

   b. 81 ______________________

   c. 125 ______________________

   d. 224 ______________________

   e. 500 ______________________

3. Circle the correct answer.

a. $140 \div 7$	30	50	20	10
b. $144 \div 4$	26	86	76	36
c. $675 \div 9$	75	35	25	15
d. $724 \div 2$	262	862	362	962
e. $3{,}192 \div 7$	356	456	256	156

4. Calculate the average score for each student. Use a calculator to help you.

Student	Test 1	Test 2	Test 3	Test 4	Average
a. Cal	86	30	59	66	
b. Taylor	92	95	90	50	
c. Reen	49	88	52	57	
d. Lois	73	72	75	60	

Name **Date**

1. Rewrite and calculate on a separate sheet of paper.

a. \$12.12 ÷ 4 **b.** \$204.12 ÷ 9 **c.** \$736 ÷ 5 **d.** \$195.60 ÷ 3 **e.** \$73 ÷ 5

2.
a. 9)920 **b.** 4)1,206 **c.** 8)819 **d.** 3)334 **e.** 8)556

f. 7)877 **g.** 6)1,343 **h.** 2)193 **i.** 4)495 **j.** 9)917

k. 6)816 **l.** 7)369 **m.** 5)283 **n.** 3)296 **o.** 7)591

3. Calculate carefully.

a. 3 L ÷ 5 = ____________________

b. 4.95 kg ÷ 6 = ____________________

c. 67.5 m ÷ 9 = ____________________

d. 216 hours = ____________________ days

e. 10 dozen ÷ 10 = ____________________

4. Complete the missing digits.

a. 4)1 0 9 = 2 7 r _

b. 6)9 1 8 = 1 _ 3

c. _)1, 1 0 7 = 3 6 9

d. 5)2, 3 7 3 = _ 7 4 r 3

5. Write *true* or *false* for each of the following.

a. 180 ÷ 3 = 60 ________ **b.** 500 ÷ 5 = 25 ________ **c.** 531 ÷ 9 = 59 r6 ________

d. 750 ÷ 6 = 125 ________ **e.** 125 ÷ 5 = 25 ________ **f.** 114 ÷ 8 = 14.6 ________

g. 300 ÷ 4 = 75 r2 ________ **h.** 435.4 ÷ 7 = 62.2 ________ **i.** 121 ÷ 2 = 468 ________

6. On a separate sheet of paper, write your own story problems for these equations.

a. 8)5 2 8 = 6 6

b. 5)1 7 6 = 3 5 r 1

c. 4)9 4 = 2 3 . 5

Name **Date**

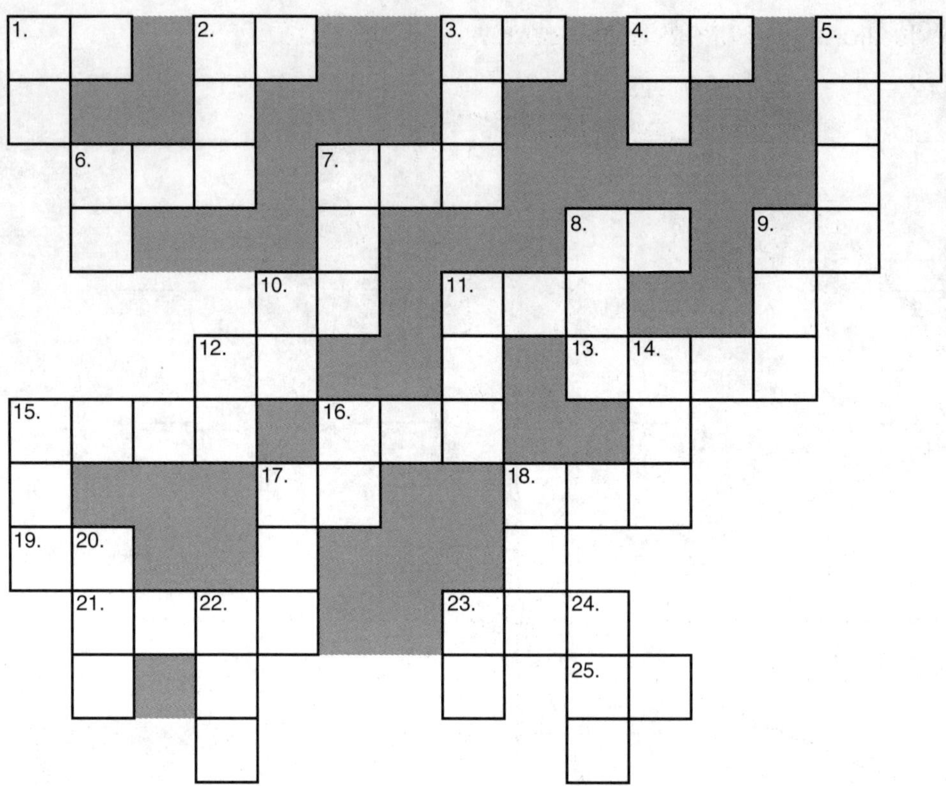

Across

1. $10\overline{)100}$
2. 216 ÷ 6
3. divide 960 by 12
4. 427 ÷ 7
5. $\sqrt{225} \div 1$
6. 3,384 ÷ 6
7. 20 centuries ÷ 4 = ___ years
8. 240 divided by 5
9. $\frac{1}{5}$ of 200
10. largest factor of 81
11. $4\overline{)868}$
12. (62 + 33) ÷ 5
13. $\frac{1}{4}$ of 10,000
15. 90,000 ÷ 10
16. quotient when 1,344 is divided by 7
17. 6 for 36¢. How many for $1.20?
18. 64 x 10 ÷ 4
19. 144 ÷ 12
21. $11\overline{)65{,}131}$
23. How many 10 cm pieces in 62.2 m?
25. 1, 3, 9, ____ , 81

Down

1. 80 ÷ 8
2. How many shares of 3 in 1,032?
3. 9 pairs of socks at 90¢ per pair
4. divide 536 by 8
5. $49.20 ÷ 4
6. $(6 + 4)^2 \div 2$
7. ____ ÷ 3 = 187
8. (472 x 2) ÷ 2
9. 4,000 ÷ 10
10. 8,900 ÷ 100
11. 1,000 ÷ 5 + 12
12. 260 ÷ 26
14. 171 ÷ 3 x 10
15. 901 ÷ 1
16. How many 12s in 120?
17. average of 125, 183, 270, 266
18. 976 divided by 8
20. $\frac{1}{4}$ of 1,000
22. $\sqrt{36} \div 3 \times 100$
23. 3 centuries ÷ 5 = _____ years
24. 1,338 legs, how many insects?

Name **Date**

Calculate the equations and check the code for the wheel of fun.

___ ___ ___   ___ ___ ___ ___ ___ ___ ___
44 59 31   26 15 31 23 44 59 82

___ ___ ___ ___   ___ ___ ___ ___ !
58 62 59 59   95 42 92 36

Name **Date**

1. Use a calculator to complete the table.

	810	560	777	403	200	640	392	930
÷ 1								
÷ 10								
÷ 100								

Can you see a pattern? Explain it to a friend.

2. Explain why you cannot divide by zero.

____

____

3. Show how multiplication and division are related. Follow the example below.

180 ÷ 3 = 60 ⟶ *3 x 60 = 180*

a. 564 ÷ 4 = 141 ⟶ ____________________

b. 999 ÷ 9 = 111 ⟶ ____________________

c. $8\overline{)1{,}608}$ = _____ ⟶ ____________________

d. $5\overline{)405}$ = _____ ⟶ ____________________

4. Which path will correctly take Chris to the skate ramp with a final answer of 23? Calculate all paths and fill in the blank spaces.

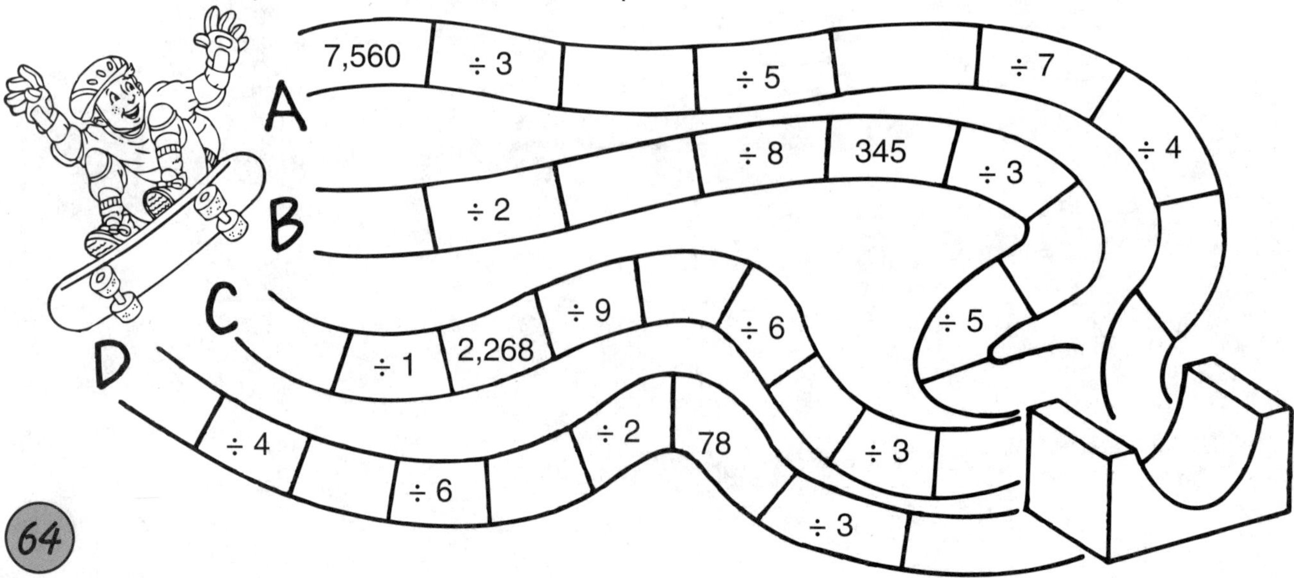

Name **Date**

1. **a.** 693 ÷ 3 **b.** 810 ÷ 10 **c.** 3,157 ÷ 7 **d.** 252 ÷ 9 **e.** 420 ÷ 5

2. **a.** (12 ÷ 6) x 2 = ____________
   b. quotient of 27 and 3 = ____________
   c. average of 27.5, 18.5, 46, 32 ____________
   d. (6 + 3 x 2 - 5) ÷ 7 = ____________
   e. Write a word problem for 364 ÷ 7. ____________

3. Calculate and number in descending order.

   a. ____ $2\overline{)3,216}$ **b.** ____ $10\overline{)19,020}$ **c.** ____ $8\overline{)1,248}$ **d.** ____ $6\overline{)3,120}$ **e.** ____ $4\overline{)3,932}$

4. Find the missing digits.

   a. 123 over __)246 **b.** 2_2 over 5)1,010 **c.** 301 over 4)1_ 0_ **d.** 61 over 6)36_ **e.** 121 over _)_47

5. True or false?

   a. 162 ÷ 6 = 28 r3 ____________ **b.** 199 ÷ 3 = 66 r1 ____________
   c. 114 ÷ 8 = 19 ____________ **d.** 279 ÷ 5 = 55 r4 ____________
   e. 190 ÷ 4 = 47 r2 ____________ **f.** 738 ÷ 9 = 82 r7 ____________
   g. 166 ÷ 2 = 83 ____________ **h.** 992 ÷ 10 = 992 ____________

6. Write a number sentence and the answer.

   a. Eight truck drivers have been contracted to deliver 768 imported cars. Each truck can hold six vehicles. How many trips will each driver take to deliver the imported cars?

   ____

   b. These cars are to be equally shared between four car dealerships. How many cars are expected to be delivered at A1 Auto Sales?

   ____

Name **Date**

Help Fred the frog reach his lily pad.

He moves only vertically or horizontally—NOT diagonally!

He can only hop onto squares that are divisible by 7, 9, or 11.

If you are brave, help him without the use of a calculator.

5,776	1,189	4,118	2,356	1,173	1,548	6,820	1,029	704	
1,073	361	899	226	1,955	266	1,615	901	1,209	897
915	1,116	6,631	4,059	1,360	5,643	1,736	5,889	745	158
713	810	901	760	584	2,410	602	585	345	6,105
5,615	1,196	799	1,411	1,222	6,760	609	1,404	1,027	319
713	2,002	801	189	1,177	986	1,612	605	884	2,957
1,020	2,429	1,380	570	3,384	522	1,099	946	625	157
1,037	737	979	1,003	2,920	1,564	530	697	539	811
578	246	623	1,131	409	818	794	376	263	7,057
121	1,792	963	219	304	1,102	4,112	2,876	3,006	914

WORKING WITH DIVISION

Unit 2

Basic division
Patterns
Estimating
Decimals
Double-digit division

Objectives

- *recall and use division facts up to 10 x 10*
- *model, explain, and continue patterns related to number facts for division*
- *select appropriate strategies to approximate and calculate solutions to problems*
- *use understood written methods to divide numbers by two-digit numbers*
- *use understood written methods to divide whole numbers, money, and measures*
- *use a calculator efficiently for operating on numbers*
- *describe situations which may be represented by given symbolically expressed number sentences*
- *estimate and calculate mentally*
- *calculate with decimals, including dividing by whole numbers up to 10*

Language

share, group, divide, divided by, division, remainder, factors, quotient, average, single-digit division, divisor, dividend, double-digit division

Materials/Resources

scratch paper, calculators

Contents of Student Pages

 * *Materials needed for each reproducible student page*

Remember

Encourage students to:

- ❑ *seek help when unsure.*
- ❑ *explain their understandings.*
- ❑ *use a Times Table Poster if needed.*
- ❑ *use concrete materials if necessary.*
- ❑ *use calculators to check their work.*

Additional Activities

- *Design a division board game.*
- *Daily division grid. Students can beat their own time if using the same grid outline. They develop accuracy and confidence.*
- *Play Dividing Heads: 4 players. Each player places a card on his/her forehead that shows a division number sentence, e.g. 28 ÷ 4. Each player in turn asks yes/no questions about their answer, e.g., "Is my number odd?" Class answers yes/no. Continue until players ask 5 questions each, then they try to guess their answer.*
- *Use supermarket advertisements and explore how division is used.*
- *Brainstorm and display division concepts.*

Answers

Page 69 Division

1. a. 154 < 315 e. 66 = 66 h. 202 = 202
   b. 431 > 314 f. 101 < 125 i. 83 < 84
   c. 87 < 181 g. 53 = 53 j. 70 > 61
   d. 16 > 15
2. a. 3, 3, 6 c. 8, 2 e. 9
   b. 2, 3 d. 4
3. a. 0.25 e. 0.1 i. 1.6
   b. 0.5 f. 0.2 j. $7.\overline{88}$
   c. 0.75 g. $0.\overline{66}$ k. 2
   d. $0.\overline{33}$ h. 2.5 l. 1
4. a. 39 b. 8
5. Answers will vary.

Page 70 Calculated Help

Errors

1. 411 9. 0.92
4. 15 12. 750,000
7. 5 15. 18

Wheels

17. a. 9, 30, 82, 22, 93, 43, 132, 15
    b. 14, 23, 190, 56, 100, 204, 77, 1

Page 71 Patterns

1. a. 6.5, 0.65 d. 4.1, 0.41 g. 7.08, 0.708
   b. 12.6, 1.26 e. 8.46, 0.846 h. 2.91, 0.291
   c. 3.7, 0.37 f. 3.15, 0.315
2. a. left
   b. one, two
   c. decimal point moves three places to the left
3. a. 2.1, 0.21 c. 9.2, 0.92
   b. 4.6, 0.46 d. 8.3, 0.83
4. The decimal point moves one place to the right when you multiply by 10 and two places to the right when you multiply by 100.

5. a. 3.3 f. 2.26
   b. 0.87 g. 9.35
   c. 0.99 h. 0.422
   d. 1.6 i. 0.073
   e. 5.4
6. Check individual work.

Page 72 Dividing Decimals

1. a. 1.2 e. 3.2 h. 1.1
   b. 3.3 f. 1.0 i. 1.1
   c. 2.1 g. 1.1 j. 1.6
   d. 1.1
2. a. 0.11 e. 0.02 i. 0.12
   b. 1.1 f. 5.3 j. 1.01
   c. 1.01 g. 1.32 k. 3.4
   d. 1.04 h. 1.02 l. 2.02
3. a. 0.9 e. 2.04 h. 0.41
   b. 0.8 f. 0.71 i. 0.31
   c. 0.2 g. 3.04 j. 0.35
   d. 0.52
4. a. 1.16 e. 1.02 h. 0.03
   b. 0.04 f. 2.27 i. 0.82
   c. 0.61 g. 11.16 j. 0.35
   d. 0.62
5. a. 4.6 m b. 56 pounds
6. a. 1.39 b. 1.25 c. 3.56

Page 73 Double-Digit Division

1. a. 50 c. 20 e. 30
   b. 30 d. 90
2. a. 16 c. 23 e. 25
   b. 46 d. 52 f. 79
3. a. 12 days b. 23 pounds
4. Check individual work.

Page 74 Assessment

1. a. 1,456 c. 1,023 e. 960
   b. 589 d. 420
2. a. 189 c. 234 e. 852
   b. 9.1 d. 3.75
3. a. 2.7 d. 0.99 g. 0.864
   b. 0.4726 e. 3.97 h. 0.531
   c. 2.101 f. 0.8 i. 0.086
4. a. 0.06 c. 3.47 e. 12.8
   b. 0.025 d. 72.3
5. a. i. $8.50
   ii. $7.50
   iii. $25
   b. 158 boxes

Name **Date**

1. Divide, then put the correct sign (>, <, or =) in the box.

a.	$8\overline{)1,232}$	☐	$3\overline{)945}$	**b.**	$2\overline{)862}$	☐	$6\overline{)1,884}$
c.	$3\overline{)261}$	☐	$9\overline{)1,629}$	**d.**	$5\overline{)80}$	☐	$9\overline{)135}$
e.	$9\overline{)594}$	☐	$4\overline{)264}$	**f.**	$3\overline{)303}$	☐	$9\overline{)1,125}$
g.	$8\overline{)424}$	☐	$5\overline{)265}$	**h.**	$7\overline{)1,414}$	☐	$6\overline{)1,212}$
i.	$4\overline{)332}$	☐	$9\overline{)756}$	**j.**	$2\overline{)140}$	☐	$8\overline{)488}$

2. What are the missing digits?

a. $\begin{array}{r} 12\square 0 \\ 3\overline{)\square\square 90} \end{array}$ **b.** $\begin{array}{r} 6\square 8 \\ 5\overline{)\square 140} \end{array}$ **c.** $\begin{array}{r} 3\square 7 \\ 6\overline{)232\square} \end{array}$ **d.** $\begin{array}{r} \square 55 \\ 7\overline{)3,185} \end{array}$ **e.** $\begin{array}{r} 236 \\ \square\overline{)2,124} \end{array}$

3. Match each fraction to its equivalent decimal.

a. $\frac{1}{4}$ 0.5

b. $\frac{1}{2}$ 0.2

c. $\frac{3}{4}$ $0.\overline{33}$

d. $\frac{1}{3}$ 0.25

e. $\frac{1}{10}$ 0.75

f. $\frac{1}{5}$ 0.1

g. $\frac{2}{3}$ $0.\overline{66}$

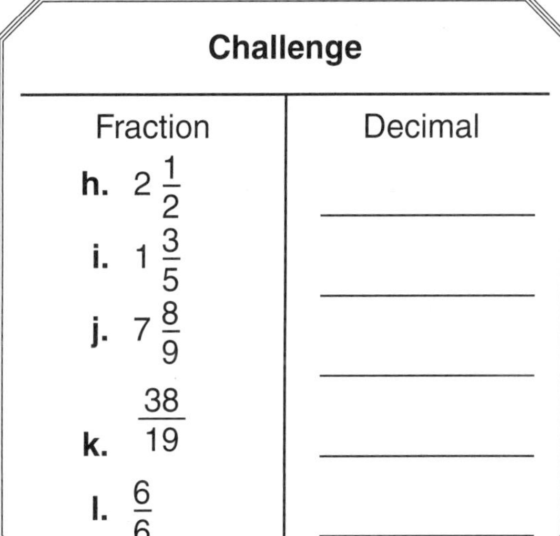

Challenge

Fraction	Decimal
h. $2\frac{1}{2}$	________
i. $1\frac{3}{5}$	________
j. $7\frac{8}{9}$	________
k. $\frac{38}{19}$	________
l. $\frac{6}{6}$	________

4. **a.** How many taxis are needed to transport 154 passengers if each taxi can hold 4 passengers? ________________

   b. 116 bread rolls were taken out of the oven. How many complete "baker's dozen" packets can be packaged? ________________

5. On a separate sheet of paper, write word problems for these division questions.

a. $\begin{array}{r} 220 \\ 8\overline{)1,760} \end{array}$ **b.** $\begin{array}{r} 635 \\ 6\overline{)3,810} \end{array}$ **c.** $\begin{array}{r} 62 \\ 9\overline{)558} \end{array}$

Name **Date**

Use a calculator to help Joshua check his division test for incorrect answers. Check the correct answers and change the incorrect answers to the right answers.

Division Test

Name: Joshua

Date: September 15th

	Answer
1. 2,877 ÷ 7	41
2. Quotient of 81 and 3.	27
3. Divide 3,760 by 8.	470
4. How many egg cartons for 170 eggs if each carton holds a dozen eggs?	14
5. Tamara traveled 204 miles in 3 hours. What was the average distance traveled per hour?	68 mph
6. 429 ÷ 2 = 214 r ______	1
7. $17.50 ÷ _____ = $3.50	0.5
8. 0.123 ÷ 3	0.041

	Answer
9. $8\overline{)7.36}$	92
10. ____ ÷ 0.5 = 22.3	11.15
11. Average of 261, 112, and 32.	135
12. Half of one-and-a-half million.	75,000
13. One thousand three hundred twenty-three divided by 9.	147
14. $\frac{(2 + 2.5 + 1.5)}{3}$	2
15. 8,370 ÷ ____ = 465	1.8
16. $\frac{6 \times 2 \times 4 \times 1}{2}$	24

17. **a.**

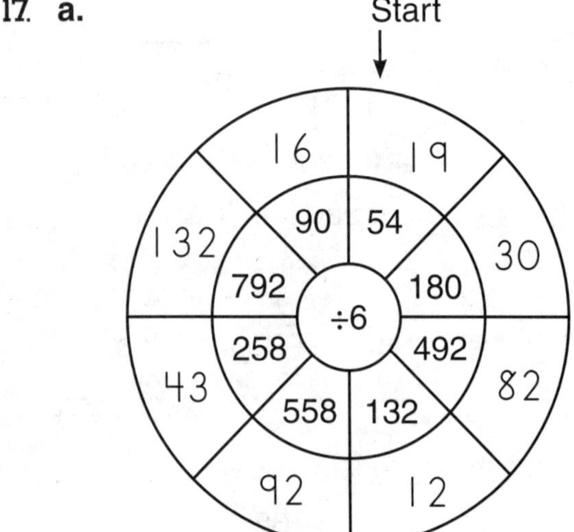

b.

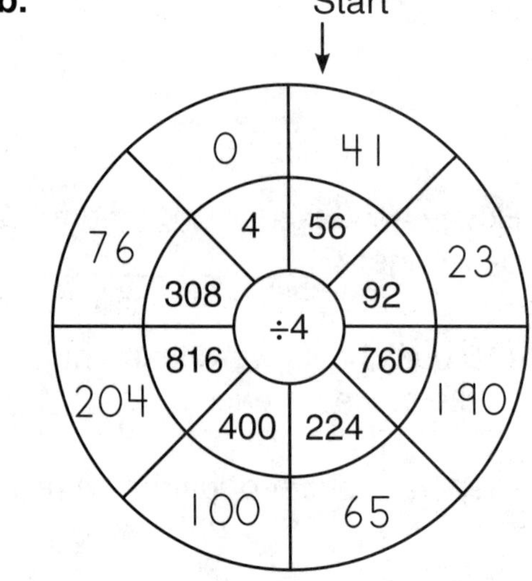

Name **Date**

1. Use a calculator to complete the table. Carefully observe the place value position of each number.

		÷ 10	÷ 100
a.	65		
b.	126		
c.	37		
d.	41		
e.	84.6		
f.	31.5		
g.	70.8		
h.	29.1		

2. With a partner discuss your answers.
   a. Which way did the decimal point move?

   ____

   b. How many places did the decimal point move when:
   - dividing by 10? _______________
   - dividing by 100? _______________

   c. What will happen when you divide by 1,000? Check your assumptions on a calculator.

   ____

3. Now that you know about patterns when dividing, quickly answer these questions.

   a. 21 ÷ 10 = _____ **b.** 46 ÷ 10 = _____ **c.** 92 ÷ 10 = _____ **d.** 83 ÷ 10 = _____

   21 ÷ 100 = _____ 46 ÷ 100 = _____ 92 ÷ 100 = _____ 83 ÷ 100 = _____

4. If you multiply the numbers in Question 1 by 10 and 100, what happens? Discuss the movement of the decimal point with your partner.

5. Try these. Check using a calculator.

   a. 33 ÷ 10 = _______ **b.** 87 ÷ 100 = _______ **c.** 99 ÷ 100 = _______

   d. 16 ÷ 10 = _______ **e.** 54 ÷ 10 = _______ **f.** 22.6 ÷ 10 = _______

   g. 93.5 ÷ 10 = _______ **h.** 42.2 ÷ 100 = _______ **i.** 0.73 ÷ 10 = _______

6. Make your own division equations and swap with your partner. Check the answers using a calculator.

   a. _______________ **b.** _______________ **c.** _______________

   d. _______________ **e.** _______________ **f.** _______________

   g. _______________ **h.** _______________ **i.** _______________

Name	Date

1. **a.** 4)4.8 **b.** 2)6.6 **c.** 4)8.4 **d.** 6)6.6 **e.** 3)9.6

   f. 9)9.0 **g.** 5)5.5 **h.** 7)7.7 **i.** 8)8.8 **j.** 6)9.6

2. **a.** 3)0.33 **b.** 5)5.5 **c.** 6)6.06 **d.** 4)4.16

   e. 8)0.16 **f.** 2)10.6 **g.** 3)3.96 **h.** 7)7.14

   i. 4)0.48 **j.** 9)9.09 **k.** 2)6.8 **l.** 5)10.10

3. **a.** 5)4.5 **b.** 8)6.4 **c.** 9)1.8 **d.** 4)2.08 **e.** 3)6.12

   f. 7)4.97 **g.** 3)9.12 **h.** 6)2.46 **i.** 5)1.55 **j.** 5)1.75

4. **a.** 2)2.32 **b.** 5)0.2 **c.** 6)3.66 **d.** 7)4.34 **e.** 3)3.06

   f. 3)6.81 **g.** 4)44.64 **h.** 9)0.27 **i.** 8)6.56 **j.** 5)1.75

5. **a.** Mrs. Edwards needs to cut six equal lengths of ribbon to wrap presents. How long will each one be if the ribbon is bought on a 27.6 meter roll? ____________________

   b. The combined weight of eight boys is 448 pounds. What is their average weight?

   ____

6. Estimate first, then use your calculator.

   a. 0.6)0.834 Est. [ ] **b.** 0.5)0.625 Est. [ ] **c.** 0.2)0.712 Est. [ ]

Name **Date**

Estimate by rounding to the nearest ten, hundred, or thousand. Some examples have been rounded below.

1,420	16	37	1,788	92	199
1,000	20	40	2,000	90	200

1. Circle the best estimate.

a.	$23\overline{)1,028}$	50	10	160	10
b.	$49\overline{)1,491}$	90	30	80	70
c.	$76\overline{)1,699}$	20	90	100	70
d.	$39\overline{)3,565}$	100	10	20	90
e.	$91\overline{)2,777}$	50	30	100	80

2. Long division shows everyone how the calculation is completed. Complete these long division equations on the back of this page. Show all work. Keep your columns straight.

```
    31
12)372
 - 36   (12 x 3)
   12
 - 12   (12 x 1)
    0
```

a. $15\overline{)240}$ **b.** $24\overline{)1,104}$ **c.** $33\overline{)759}$

d. $16\overline{)832}$ **e.** $13\overline{)325}$ **f.** $21\overline{)1,659}$

3. Estimate, then solve the problems. **Show Work**

a. How many days in 288 hours?

Estimate ____________ Answer ____________

b. How many pounds in 368 ounces?

Estimate ____________ Answer ____________

4. On a separate sheet of paper, write word problems for these division problems.

a. $17\overline{)187}$ **b.** $32\overline{)672}$ **c.** $25\overline{)450}$

Name **Date**

1. **a.** $6\overline{)8,736}$ **b.** $2\overline{)1,178}$ **c.** $9\overline{)9,207}$ **d.** $5\overline{)2,100}$ **e.** $3\overline{)2,880}$

2. Rewrite and calculate below to find the answers.

a. 1,323 ÷ 7	**b.** $\frac{27.3}{3}$	**c.** 1,872 ÷ 8	**d.** Write $3\frac{3}{4}$ as a decimal.	**e.** $\frac{3,408}{12 \div 3}$

3. Watch the decimal point!
   a. 27 ÷ 10 = ________ **b.** 47.26 ÷ 100 = ________ **c.** 210.1 ÷ 100 = ________
   d. 99 ÷ 100 = ________ **e.** 39.7 ÷ 10 = ________ **f.** 8 ÷ 10 = ________
   g. 86.4 ÷ 100 = ________ **h.** 531 ÷ 1,000 = ________ **i.** 0.86 ÷ 10 = ________

4. Divide showing all work.

   a. 0.36 ÷ 6 **b.** 0.125 ÷ 5 **c.** 20.82 ÷ 6 **d.** 289.2 ÷ 4 **e.** 102.4 ÷ 8

5. Show your work on the back of this page.

   a. Jane paid $21.25 for 2.5 pounds of perch fillets.

   i. How much was the fish per pound? ________

   ii. What would the change be from $50 if she bought 5 pounds of fish? ________

   iii. If the price of perch fillets rose by $1.50 per pound, how much will she pay for 2.5 pounds next week? ________

   b. The local charity group sold biscuits at $3.50 per box. In two days they had collected $553.00. How many boxes were sold? ________

NUMBER FACTS

Quick recall of number facts is an essential skill to develop for all work in mathematics. It allows for accuracy, speed, and good estimation. In these units, this recall is encouraged by exercises with factors and multiples; square numbers and square roots; addition and subtraction facts; tests for divisibility; prime and composite numbers; order of operations; and products, sum, and difference.

Further practice is gained by completing patterns, filling in grids, division wheels, and magic squares. The activity page is a large crossnumber puzzle that requires students to use all number facts to find the palindromic answers. Two assessment pages are included.

UNDERSTANDING NUMBER FACTS

Unit 1

Factors
Four processes
Order of operations
Terms and expressions

Objectives

- *automatically recall and use multiplication and division facts to 10 x 10*
- *use number skills involving whole numbers to solve problems*
- *estimate and calculate mentally, including adding and subtracting two-digit numbers*
- *use various combinations of numbers to complete number sentences*
- *use understanding of numbers and number relationships to construct and complete statements of equality*

Language

multiplication, factor, product, share, group division, quotient, remainders, addition, addends, total, sum, subtract, minus, difference, less, grouping, symbols, magic squares, plot, even, odd

Materials/Resources

highlighters, scratch paper, calculators

Contents of Student Pages

* *Materials needed for each reproducible student page*

Page 78 Multiplication
using the terms "factors" and "product"

Page 79 Using Known Sums
using the terms "total" and "sum"; short cuts utilizing known combinations

Page 80 Subtraction
by counting on and counting back; using the terms "difference," "minus," "less than"; as the reverse of addition

Page 81 Order of Operations
mixed examples
* *scratch paper*

Page 82 Revision Puzzles
solving different types of math puzzles
* *highlighters*

Page 83 Assessment
* *highlighters*

Page 84 Palindromes Galore Activity
* *calculators*

Remember

- ❑ *Use terms for each operation regularly in the classroom, e.g. "sum" and "total" for addition, etc.*

Additional Activities

- *There are many games to play to reinforce number facts, e.g., quizzes, team competitions, 20 Questions, Bingo, Memory, etc.*
- *Daily, one-minute tests with time or score targets keep students sharp.*

Answers

Page 78 Multiplication

1. a. 28, 63, 35, 70, 21, 49, 0, 7, 14, 42
   b. 56, 32, 16, 0, 64, 80, 40, 72, 8, 48
   c. 4, 8, 9, 0, 6
2. a. 10, 12, 120; 10, 8, 80; 7, 8, 56; 10, 5, 50; 6, 8, 48; 7, 6, 42; 9, 7, 63; 9, 9, 81; 2, 9, 18
   b. 7, 54, 45, 35, 13, 9, 7, 3, 4, 72

Page 79 Using Known Sums

1. a. 17 e. 19 i. 20
   b. 14 f. 15 j. 18
   c. 19 g. 21 k. 22
   d. 19 h. 20 l. 27
2. a. 17 e. 19 i. 30
   b. 29 f. 29 j. 26
   c. 26 g. 29 k. 18
   d. 19 h. 27 l. 24
3. Check individual work.

Page 80 Subtraction

1. a. 6 d. 9 g. 9 j. 5
   b. 3 e. 6 h. 5
   c. 7 f. 9 i. 6
2. a. 37 e. 20 h. 12
   b. 62 f. 46 i. 53
   c. 28 g. 20 j. 47
   d. 33
3. a. 7 e. 11 h. 57
   b. 7 f. 72 i. 126
   c. 9 g. 67 j. 93
   d. 15
4. a. 33 c. 22 e. 70
   b. 13 d. 30 f. 72
5. a. 62, 59, 56, 53
   b. 7, 37, 30, 23, 16
   c. 37, 29, 21, 13, 5
   d. 4, 19, 15, 11, 7
   e. 26, 20, 14, 8, 2
   f. 9, 61, 43, 34, 25

Page 81 Order of Operations

1. a. 22 d. 9 g. 44
   b. 29 e. 20 h. 3
   c. 40 f. 14 i. 40
2. a. 47 d. 39 g. 13
   b. 47 e. 26 h. 41
   c. 7 f. 45 i. 3
3. a. 72 e. 80 i. 12
   b. 73 f. 4 j. 210
   c. 0 g. 95
   d. 80 h. 45
4. a. 49 d. 1
   b. 4 e. 64
   c. 114 f. 0

Page 82 Revision Puzzles

1. Letter formed is a Z.
2. Check individual work.
3. Missing numbers by row:
   a. 15, 12, 9, 11, 2
   b. 18, 7, 10, 12, 14, 8, 6
   c. 28, 8, 14, 12, 10, 4, 6

Page 83 Assessment

1.

147	5	11	29	33
41	163	196	38	27
16	24	75	60	45
32	13	89	40	53
1	11	4	35	41
21	5	44	62	8

2. a. 2, 6, 14, 18
   b. 16, 6, 8, 13, 15, 5
   c. 22, 12, 15, 17, 16, 19, 18
3.

17	18	20	35	19	6	18
1	12	4	15	15	6	10
72	45	96	250	34	0	56

Page 84 Palindromes Galore Activity

Across

1. 339,933 17. 7,117
5. 525 19. 868
7. 37,573 21. 4,334
9. 6,666 22. 515
10. 2,552 23. 78,587
11. 77 24. 72,327
13. 33 26. 454
15. 40,304 27. 12,721

Down

1. 37,673 14. 373
2. 9,669 15. 414
3. 33 16. 48,184
4. 37,273 18. 13,631
5. 575 20. 65,756
6. 232 21. 4,774
8. 55 22. 55
11. 707 23. 777
12. 888 25. 22

Name **Date**

Product: the result of the process of multiplication
Factors: the numerals which when multiplied together, give a **product**
The **product 45** has **factors 9 and 5**.
Multiplying the **factors 4 and 8** results in the **product 32**.

1. Complete the tables.

a.

Factor	Factor	Product
7	4	
9	7	
7	5	
7	10	
3	7	
7	7	
7	0	
1	7	
7	2	
6	7	

b.

Factor	Factor	Product
7	8	
8	4	
2	8	
0	8	
8	8	
8	10	
5	8	
8	9	
8	1	
8	6	

c.

Factor	Factor	Product
9		36
9		72
7		63
9		0
6		36

2. **a.** Link two factors with their product.

Factor	Product	Factor
9	54	12
10	120	6
10	80	8
7	56	6
8	50	5
10	48	8
7	42	6
9	63	7
2	81	9
9	18	9

b. Find the missing factor or product in each set.

F	F	P
3		21
5	9	
2		26
7		49
	8	32

F	F	P
9	6	
7	5	
	4	36
8		24
9	8	

Name **Date**

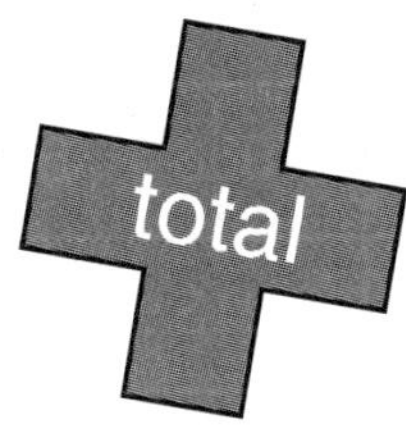

1. Use sums of ten to find these totals quickly.

a.	b.	c.	d.	e.	f.
6 7 10 + 4	4 3 + 7	8 9 + 2	2 8 + 9	9 5 + 5	6 5 + 4

g.	h.	i.	j.	k.	l.
9 7 1 + 4	2 8 6 + 4	8 5 5 + 2	5 9 3 + 1	7 3 6 + 6	3 8 7 + 9

2. Use well-known sums to make adding quicker (e.g., 6 + 6 + 8 + 6 = 12 + 14 = 26).

a.	b.	c.	d.	e.	f.
7 4 4 + 2	9 9 2 + 9	1 9 8 + 8	8 4 3 + 4	5 3 8 + 3	4 6 10 + 9

g.	h.	i.	j.	k.	l.
15 5 + 9	12 7 + 8	10 9 + 11	11 9 + 6	9 3 + 6	4 12 + 8

3. Circle groups of three addends that total 20.

15	8	10	4	0	12	9	16	13	7	11	3
2	3	2	9	8	5	6	4	2	9	9	4
5	1	11	7	7	14	20	3	6	4	13	0

Name **Date**

Counting On: 75 – 69 = [6] 69 and 1 = 70 and 5 more equals 75. 1 and 5 = 6

1. Count on to answer the following:

a. 54 – 48 = _____ **b.** 30 – 27 = _____ **c.** 63 – 56 = _____ **d.** 86 – 77 = _____

e. 55 – 49 = _____ **f.** 68 – 59 = _____ **g.** 77 – 68 = _____ **h.** 23 – 18 = _____

i. 50 – 44 = _____ **j.** 92 – 87 = _____

Counting Back: 54 – 13 = [41] Count back 10 = 44. Count back 3 more = 41.

2. Count back to answer the following:

a. 45 – 8 = _____ **b.** 66 – 4 = _____ **c.** 36 – 8 = _____ **d.** 77 – 44 = _____

e. 35 – 15 = _____ **f.** 58 – 12 = _____ **g.** 39 – 19 = _____ **h.** 26 – 14 = _____

i. 65 – 12 = _____ **j.** 90 – 43 = _____

Subtraction as the reverse of addition: 28 – 13 = 15 and 15 + 13 = 28

3. Subtract to answer the following:

a. 60 – _____ = 53 **b.** 35 – _____ = 28 **c.** 48 – _____ = 39 **d.** 40 – _____ = 25

e. 29 – _____ = 18 **f.** _____ – 15 = 57 **g.** _____ – 22 = 45 **h.** _____ – 18 = 39

i. _____ – 41 = 85 **j.** _____ – 9 = 84

4. **a.** 49 subtract 16 = _____ **b.** Difference between 62 and 49 = _____

c. 70 less than 92 = _____ **d.** 55 minus 25 = _____

e. Take 28 from _____ to leave 42. **f.** Difference between 13 and 85 = _____

5. Complete these subtraction ladders.

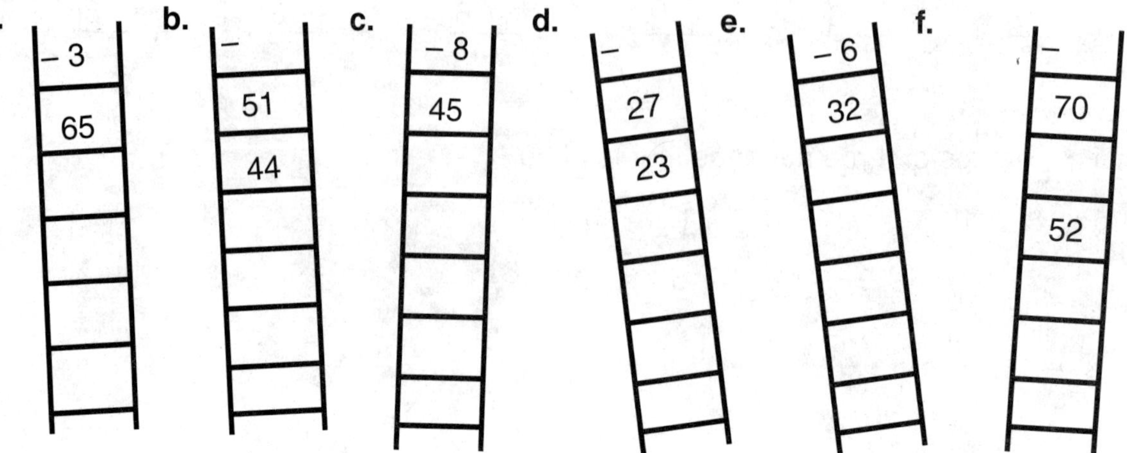

Name **Date**

Order of Operations

1. Parentheses and Exponents

2. Multiplication and Division from left to right

3. Addition and Subtraction in the order that they appear

$$\begin{aligned} 12 - 4 + (5 \times 2) + 7 \times 4 &= 12 - 4 + 10 + 7 \times 4 \\ &= 12 - 4 + 10 + 28 \\ &= 46 \end{aligned}$$

Set these out properly on a separate sheet of paper. Write the answers on this page.

1. **a.** (3 x 5) + 9 – 2 = ______ **b.** 7 + (9 x 3) – 3 – 2 = ___ **c.** 12 – 4 + (8 x 4) = ______
   d. 8 + (15 ÷ 3) – 4 = ______ **e.** 16 – 12 ÷ 4 + 7 = _____ **f.** 17 + 4 – 28 ÷ 4 = ______
   g. 6 x (5 + 4) – 10 = ______ **h.** (20 + 4) ÷ 6 – 1 = _____ **i.** 36 + 7 – (18 ÷ 6) = _____

2. **a.** 50 – 15 ÷ (3 + 2) = _____ **b.** 50 – 15 ÷ 3 + 2 = _____ **c.** (50 – 15) ÷ (3 + 2) = __
   d. 30 ÷ 6 + 4 x 8 + 2 = ____ **e.** 30 ÷ (6 + 4) x 8 + 2 = __ **f.** 30 ÷ 6 + 4 x (8 + 2) = __
   g. 25 + 14 – 9 x 3 + 1 = ___ **h.** 25 + (14 – 9) x 3 + 1 = _ **i.** 25 + 14 – 9 x (3 + 1) = _

3. **a.** 12 x (9 + 5 – 8) = ______ **b.** (8 x 9 + 4) – 12 ÷ 4 = __ **c.** (15 ÷ 15 x 2) – 2 = _____
   d. 100 – (3 x 6 + 2) = _____ **e.** 50 + (4 + 6) x 3 = _____ **f.** (3 x 4 x 2) – (5 x 4) = __
   g. (7 + 8) x (35 ÷ 5) – 10 = __ **h.** 5 x 9 – 3 x 9 + 4 x 9 – 9 = ________
   i. 150 ÷ (2 x 15) + 7 = ____ **j.** 200 + (10 x 5) ÷ 5 = __________

4. **a.** 60 – 5 – 5 – 5 ÷ 5 = ________ **b.** 3 x 1 x 1 x 1 + 1 = ________
   c. 15 + 10 x 10 – 10 ÷ 10 = ________ **d.** 36 ÷ 3 ÷ 3 – 3 = ________
   e. 80 x (8 – 8) + (8 x 8) = ________ **f.** (4 + 6 – 3 + 9 x 2) x 0 = ________

Name	**Date**

1. Highlight all the expressions that equal 72. What letter is colored?

100 – 42	9 x 4 x 2	92 – 20	81 – 9	57 + 15	100 – 28	94 – 12
3 x 17	54 ÷ 9	96 – 18	18 x 3	4 x 2 x 5	4 x 2 x 9	36 x 3
162 ÷ 2	16 x 4	14 x 5	100 – 16	2 x 2 x 2 x 3 x 3	14 x 4	13 x 3 + 10
54 + 34	9 x 5 x 2	155 ÷ 5	2 x 9 x 4	19 x 4	24 + 9 x 2	164 ÷ 2
80 – 18	9 x 11 – 9	3 x 3 x 8	65 + 8	27 + 54	30 + 45	16 x 3 + 18
62 + 20	144 ÷ 2	55 + 27	99 – 6	4 x 17	15 x 3 + 12	9 x 4 x 3
8 x 7	2 x 35 + 2	91 – 19	3 x 24	3 x 8 + 48	3 x 4 + 60	170 ÷ 3

2. Highlight the camps in ascending order to plot the course taken by Colonel Plod across the Stony Desert.

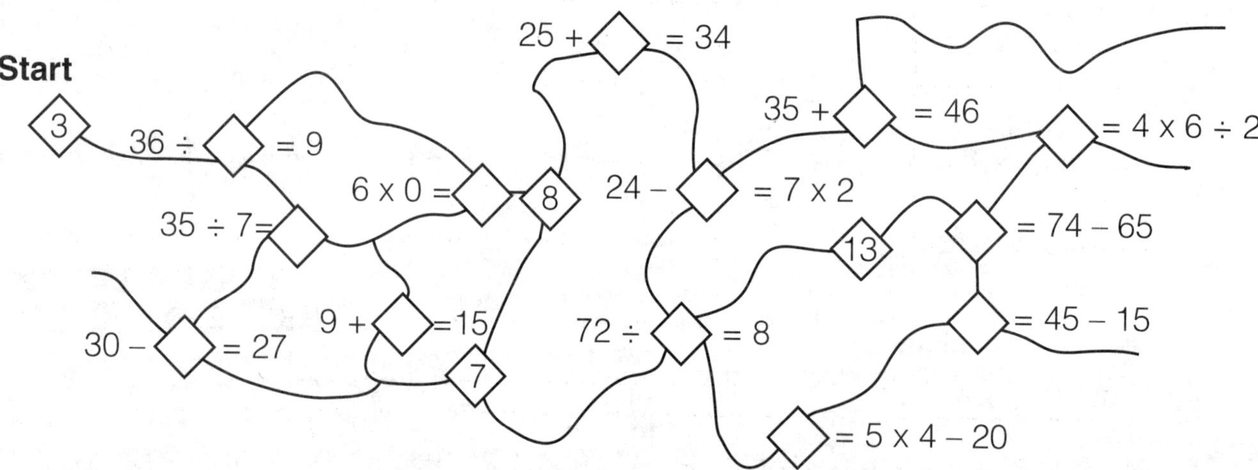

3. Complete these magic squares.

a.

1	14		4
	7	6	
8		10	5
13		3	16

b.

4	17		
15		9	
11		13	
16	5		19

c.

2		30	
24			18
16	22	20	
26			32

Name	**Date**

1. Complete the following equations, then highlight the even numbers path from **Start** to **Finish**.

(3 x 7) x 7 =	45 ÷ (3 x 3) =	23 – (2 x 6) =	36 – 7 =	22 + 7 + 4 =
26 + 15 =	100 + (7 x 9) =	200 – 100 ÷ 25 =	95 ÷ 5 x 2 =	3 x 3 x 3 =
7 + 9 =	48 ÷ 6 x 3 =	5 x 10 + 5 x 5 =	13 + 47 =	19 + 26 =
Start 45 – 13 =	81 ÷ 9 + 4 =	100 ÷ 1 – 11 =	4 x 7 + 12 =	64 – 11 =
1 x 1 =	15 – 4 =	72 ÷ 9 ÷ 2 =	5 + 9 + 16 + 5 =	4 x 4 + 5 x 5 =
17 + 4 =	35 ÷ (4 + 3) =	5 x 5 + 19 =	74 – 12 =	**Finish** (81 – 9) ÷ 9 =

2. Complete these magic squares.

a.

16		12
	10	
8		4

b.

3		17	
14	9		11
10		12	7
	4		18

c.

9		23	
20		14	
			13
21	10	11	24

3. For each pair of numbers, find the sum, difference, and product.

	8, 9	3, 15	12, 8	10, 25	2, 17	6, 0	4, 14
Sum							
Difference							
Product							

Name **Date**

Every answer in this crossnumber puzzle is a palindrome (number that can be read the same forward and backward). Use a calculator to help you complete the puzzle.

1.		2.		3.	4.		5.	6.	
				7.		8.			
9.					10.				
				11.					12.
13.	14.		15.				16.		
	17.	18.					19.	20.	
21.						22.			
				23.					
24.			25.				26.		
		27.							

Across

1. $583^2 + (100 - 7 \times 8)$
5. $5^4 - 10^2$
7. $\frac{3}{4}$ of 57,524 $- \frac{1}{2}$ of 11,140
9. $2 \times 3 \times 11 \times 101$
10. $\sqrt{64} \times (500 - 181)$
11. $\frac{1}{20}$ of 1,540
13. $\frac{3}{11}$ of 121
15. $229 \times \sqrt{256} \times 11$
17. $\frac{1}{2}$ of 8,008 $+ \frac{1}{3}$ of 9,339
19. $(279 \div 9) \times (252 \div 18) \times \sqrt{4}$
21. $(39 \times 10 + 4) \times 11$
22. $17 + 209 + 7 + 282$
23. $\frac{7}{8}$ of 100,000 $-$ 8,913
24. $(60{,}000 - 35{,}891) \times 3$
26. Double $(1{,}589 \div 7)$
27. $5{,}942 + 17{,}803 - 11{,}024$

Down

1. $331 \times 96 + 5{,}897$
2. $\frac{1}{2}$ of 16,426 $+ \frac{1}{3}$ of 4,368
3. $\sqrt{1{,}089}$
4. $50{,}000 - (30{,}000 - 17{,}273)$
5. 23×5^2
6. $2^3 \times 29$
8. $\sqrt{10{,}000} - (3^2 \times 5)$
11. $7 \times (4^3 + 37)$
12. $37 \times 2^3 \times 3$
14. $7^2 + 18^2$
15. $\frac{2}{5}$ of 1,035
16. $19 \times (185 + 132) \times 2^3$
18. 317×43
20. $23{,}945 + 917 + 31{,}519 + 9{,}375$
21. $10{,}742 - 5{,}968$
22. $119 - 4^3$
23. $6{,}993 \div \sqrt{81}$
25. $\sqrt{4} \times \sqrt{121}$

WORKING WITH NUMBER FACTS

Unit 2

Multiples
Divisibility
Squares and cubes
Triangular numbers
Patterns
Terms and expressions

Objectives

- *automatically recall and use multiplication and division facts*
- *use number skills involving whole numbers to solve problems*
- *choose and apply strategies to calculate mentally and estimate in varying contexts*
- *use various combinations of numbers to complete number sentences*
- *use understanding of numbers and number relationships to construct and complete statements of equality*

Language

multiple, factor, factor tree, product, digit pattern, prime, composite, exponents, square numbers, square root, cubic numbers, cube root, triangular numbers, length, breadth, height, volume, addend, divisibility, relationship, total, sum, subtract, quotient, minus, difference, less

Materials/Resources

calculators, scratch paper

Contents of Student Pages

* *Materials needed for each reproducible student page*

Remember

- ❑ *Use terms for each operation regularly in the classroom, e.g., "sum" and "total" for addition, etc.*

Additional Activities

- *There are many games to play to reinforce Number Facts, e.g., quizzes, team competitions, 20 Questions, Bingo, Memory, Dominoes, and timed challenges to pep up the learning of number facts.*
- *Daily time-limited activities with time or score targets keep able students sharp.*

Answers

Page 87 Multiplication Patterns

1. a. 6; 6, 2, 8, 4, 0
   b. 3; 3, 6, 9, 2, 5, 8, 1, 4, 7, 0
   c. 5; 5, 0, 5, 0
   d. 9; 9, 8, 7, 6, 5, 4, 3, 2, 1, 0
   e. 4; 4, 8, 2, 6, 0
   f. 10; 0
   g. 7; 7, 4, 1, 8, 5, 2, 9, 6, 3, 0
2. a. $5 \times 2 \times 2 \times 2$; 5×2^3
   b. $3 \times 3 \times 3 \times 2$; $3^3 \times 2$
   c. $3 \times 2 \times 3 \times 2$; $3^2 \times 2^2$
   d. $7 \times 2 \times 2$; 7×2^2

Page 88 Square Numbers

1. a. 1, 4, 9, 16, 25, 36, 49, 64, 81, 100
   b. Check individual work.
2. 121, 144, 169, 196, 225
3. a. F
   b. T
   c. T
   d. F
   e. F
   f. T
   g. T
4. a. 20
   b. 17 ft.

Page 89 Cubic and Triangular Numbers

1. a. 1, 8, 27
   b. F
   c. F
2. a. 4,096
   b. 15,625
   c. 3,375
   d. 1,728
3. 512 $in.^3$
4. a. 3
   b. 5
5. 6 cm
6. 10
7. a. 21, 28, 36, 45
   b. 91

Page 90 Tests for Divisibility

1. 9: 180; 423; 5,643; 486; 1,422; 3,501
   4: 532; 264; 176; 1,572
2. 168, 120, 72, 24, 144, 96, 264, 192
3. 18, 36, 54, 72, 90

Page 91 Revision

1. a. 72 81; 88 97; 200 209
   b. 81 69; 196 184; 400 388; 256 244
   c. 8 2; 64 16; 216 54; 1,000 250
2. a.

36	72	99	81	27	108	54	90	63	45
47	83	110	92	38	119	65	101	74	56
45	81	108	90	36	117	63	99	72	54
5	9	12	10	4	13	7	11	8	6

   b. Bottom row is one more than the top row.
   c. We added 9 then ÷ 9, which results in one more than the original number.
3. 9 (3, 1, 9), 25 (5, 1, 25), 49 (7, 1, 49)
4. 72, 60, 12, 31, 28, 4, 64, 75

Page 92 Assessment

1. a. 90
   b. 168
   c. 200
   d. 10
   e. 1,110
   f. 900
   g. 9 + 8 + 7 + 6 + 5 + 4 + 3 + 2 + 1 = 45
2. $2 \times 2 \times 2 \times 3 \times 2$; $2^4 \times 3$
3. 56 68; 12 108; 8 64; 6 48; 200 212
4. Div. by 9: 432, 126, 693, 252, 774
   Div. by 4: 432, 252, 824, 516
   Div. by 6: 432, 126, 252, 774, 516
5. a. 30
   b. 38
   c. 16
   d. 49
   e. 5
   f. 83
   g. 9
   h. yes
   i. 17
   j. 72

Name **Date**

The example in the table below shows the first five multiples of 8.
The digits in the ones place form the digit pattern (8, 6, 4, 2, 0).
These digits will repeat as you continue the list of multiples, e.g., 8 x 6 = 48.
The 8 in the ones place starts the pattern again.

1. Identify each set of multiples and write their digit pattern.

	Multiples of	Digit Pattern
Example: 8, 16, 24, 32, 40	8	8, 6, 4, 2, 0
a. 36, 42, 48, 54, 60		
b. 33, 36, 39, 42, 45, 48, 51, 54, 57, 60		
c. 25, 30, 35, 40		
d. 99, 108, 117, 126, 135, 144, 153, 162, 171		
e. 24, 28, 32, 36, 40		
f. 50, 60, 70, 80, 90		
g. 77, 84, 91, 98, 105, 112, 119, 126, 133, 140		

2. A factor tree describes a number as the product of its prime factors. Complete the factor trees below. An example has been provided for you.

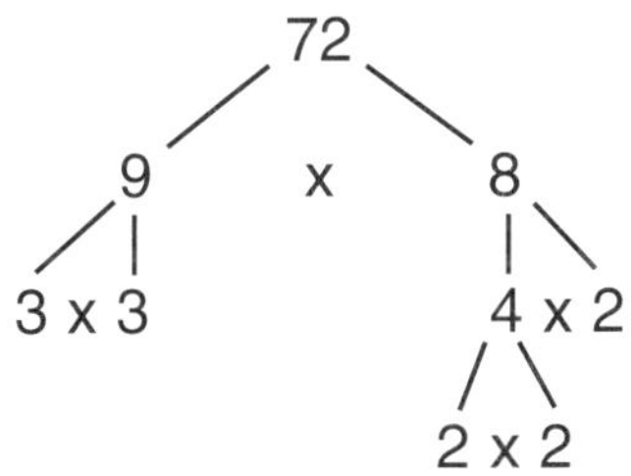

72 = 3 x 3 x 2 x 2 x 2

72 = $3^2 \times 2^3$

a.

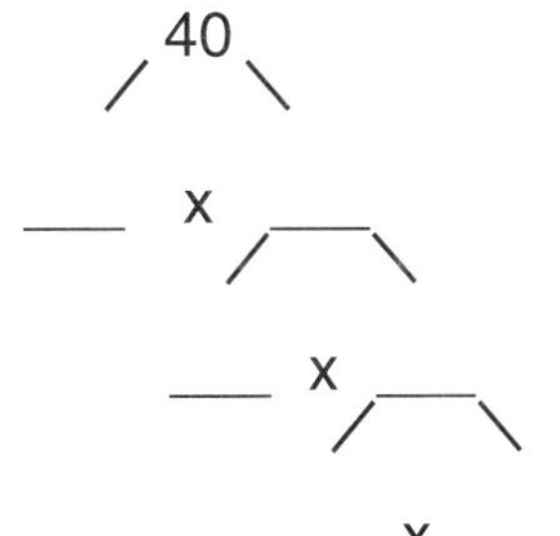

40 = ________________

40 = ________________

b.

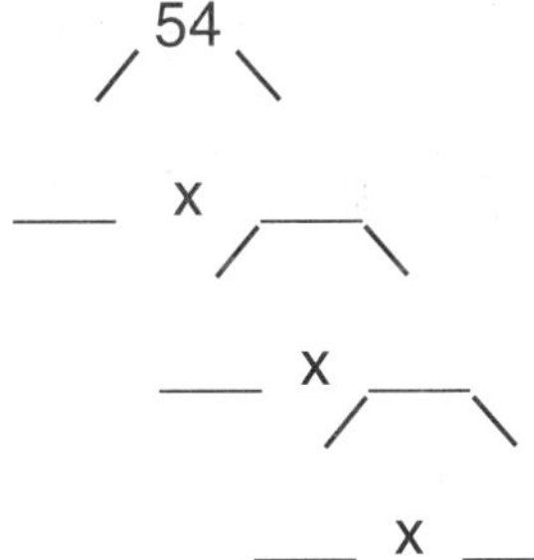

54 = ________________

54 = ________________

c.

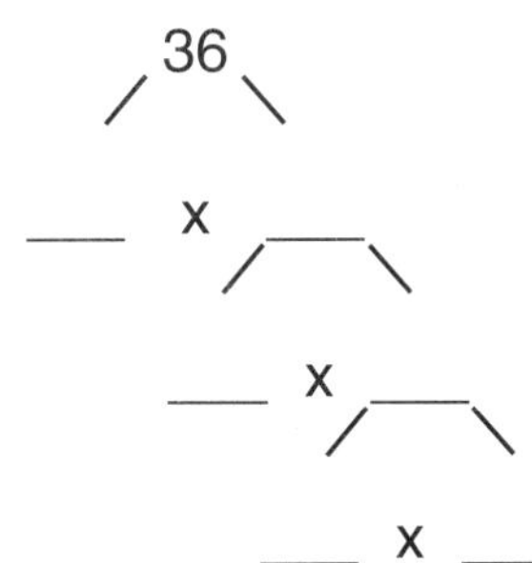

36 = ________________

36 = ________________

d. 28

___ x ___

___ x ___

28 = ________________

28 = ________________

Name **Date**

When a number is squared, it is multiplied by itself once. 10^2 = 10 x 10 = 100

1. Complete the multiplication problems below to find out the first ten square numbers. Then shade in as many as you can on the grid to show that they are true square numbers. The first two have been shaded for you.

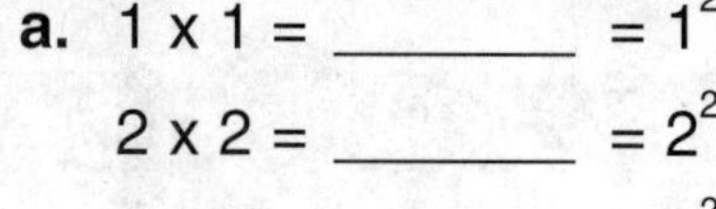

a. 1 x 1 = ________ = 1^2
2 x 2 = ________ = 2^2
3 x 3 = ________ = 3^2
4 x 4 = ________ = 4^2
5 x 5 = ________ = 5^2
6 x 6 = ________ = 6^2
7 x 7 = ________ = 7^2
8 x 8 = ________ = 8^2
9 x 9 = ________ = 9^2
10 x 10 = ____ = 10^2

b. 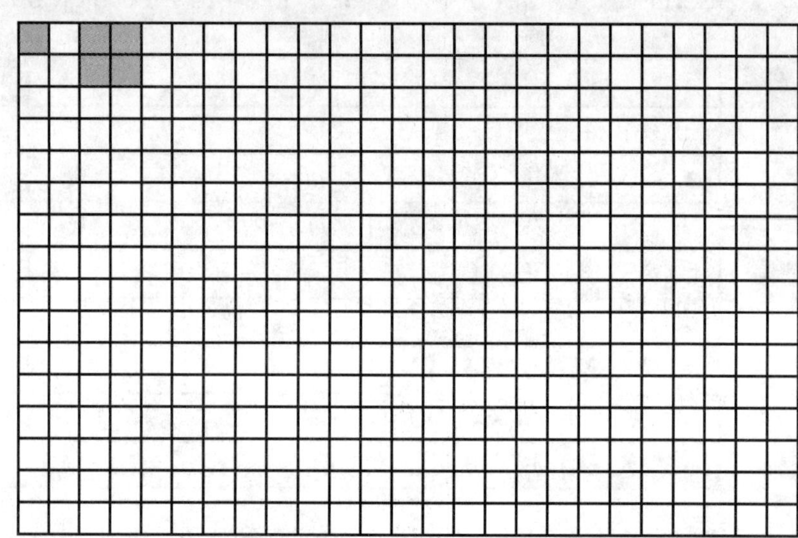

2. What are the next 5 square numbers?

____

3. True or false?

a. 4^2 is the same as 4 x 2. __________

b. The area of a 16 cm square is 256 cm^2. __________

c. 81 is the largest square number less than 100. __________

d. 1,000 is a square number. (Hint: Guess and check.) __________

e. The sum of two square numbers is always an even number. __________

f. Square numbers are a group of composite numbers. __________

g. In the series 64, 81, 100, ____ , ____ the missing numbers are 121 and 144. _____

The square root of 100 is 10. $\sqrt{100} = 10$

4. **a.** What is the square root of 400? __________

b. If the area of a square is 289 $ft.^2$, what is the length of each side? __________

Name **Date**

When a number is cubed, it is multiplied by itself twice. 4 cubed = 4^3 = 4 x 4 x 4 = 64

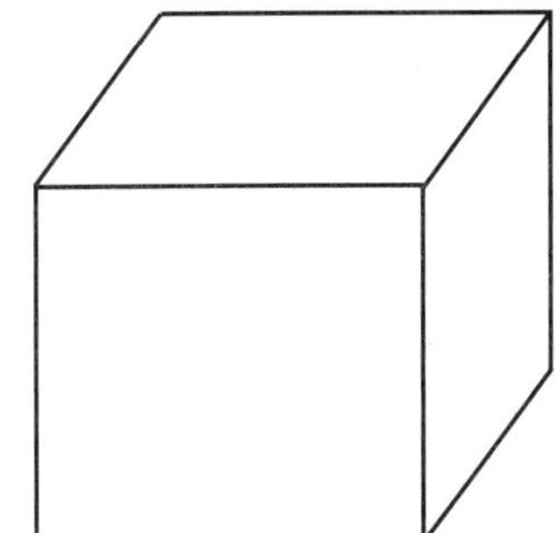

1. **a.** What are the first three cubic numbers?

   1^3 = ________ 2^3 = ________ 3^3 = ________

   True or false?

   b. 10^3 = 110 ________

   c. 7^3 is less than 100 = ________

2. Use a calculator to solve the following.

   a. 16^3 = ________ **b.** 25^3 = ________ **c.** 15^3 = ________ **d.** 12^3 = ________

Length x Width x Height = Volume

3. What is the volume of a plastic cube with sides of 8 in.? ____________

The cube root of 8 is 2. $\sqrt[3]{8} = 2$

4. What is the cube root of the following?

   a. 27 = ________ **b.** 125 = ________

5. If the volume of a plastic cube is 216 cm^3, what are its dimensions? ________

Triangular Numbers

As these billiard balls are caught in the pockets, they are forming triangular numbers.

6. Complete the fourth number and label it.

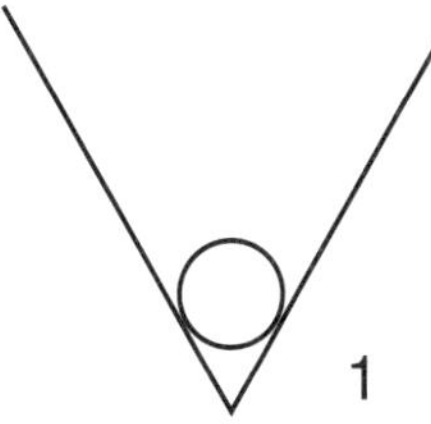

1

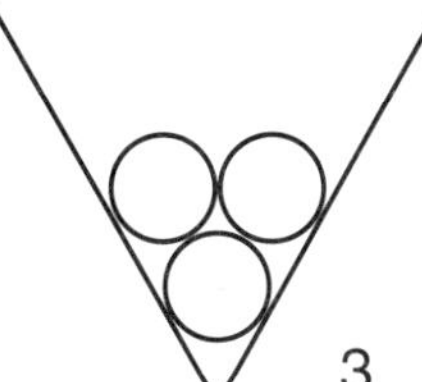

3

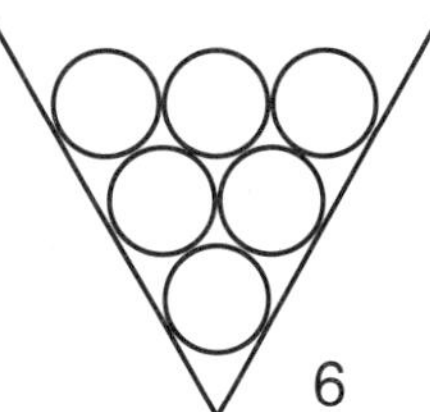

6

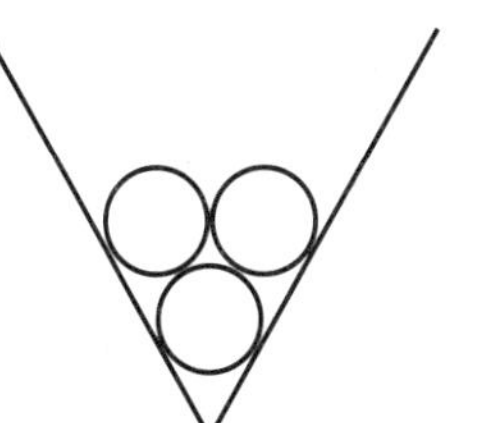

____

7. Triangular numbers are formed by adding one to the next addend.

   1 + 2 + 3 + 4 + 5 = *15 is a triangular number*

   a. What are the next four triangular numbers after 15? _____ , _____ , _____ , _____

   b. Use a calculator to find the largest triangular number less than 100. _____

Name **Date**

To see if a number can be divided equally, apply the tests of divisibility.

Factor	Test of Divisibility	Example
3	The sum of digits is divisible by 3.	381; **3 + 8 + 1 = 12**
4	The last two digits form a numeral that is divisible by 4.	14,7**32**; **32 is divisible by 4**
5	The last digit is 5 or 0.	5**0**, 23**5**, 71**5**, 1**5**, 16**0**
6	An even number whose sum of its digits is divisible by 3.	852 (even), **8 + 5 + 2 = 15**
8	The last three digits form a number that is divisible by 8.	13,**264; 264 is divisible by 8**
9	The sum of the digits is divisible by 9.	3,645; **3 + 6 + 4 + 5 = 18**
10	The last digit is 0.	23,67**0**, 2,05**0**

1. Sort the following into their correct boxes.

180 532 423 5,643 264 1,422 176 3,501 1,572 486

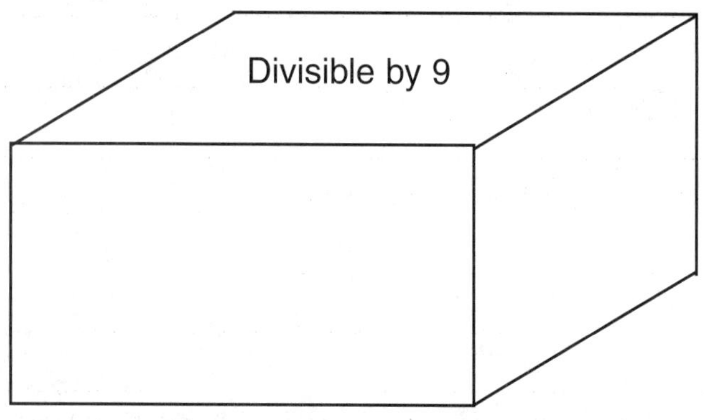

Divisible by 4

2. Which of these multiples are divisible by both 3 and 8? Circle them.

168 96 72 264
458 192 120 752
24 906 176 320 144

3. On a separate paper, write the multiples of 9 and of 6 that are less than 100. Circle those that are common to both groups. Write them here.

____

These numbers are called *common multiples* of 9 and 6.

Name **Date**

1. Calculate the missing numbers. Follow the arrows to help you.

a.

n	n x 8	n x 8 + 9
6	48	57
9		
11		
25		

b.

a	a^2	$a^2 - 12$
9		
14		
20		
16		

c.

m	m^3	$m^3 \div 4$
2		
4		
6		
10		

2. **a.** Starting at the top of each column, complete steps 1–4 to finish the table. The first column has been done for you. Follow the arrows to help you.

		2	4	8	11	9	3	12	6	10	7	5
Step 1	x 9	18										
Step 2	+ 11	29										
Step 3	– 2	27										
Step 4	÷ 9	3										

b. What is the relationship between the top row and the bottom row? ______________

____

c. Why is this so? ______________________________

____

3. The number 14 has four factors, 7, 2, 14, and 1. Circle the following numbers that have only three factors.

20, 36, 9, 18, 81, 25, 99, 49

4. Complete the following flowchart.

Start 8

x 9 = → _____ – 12 =

_____ ÷ 5 = → _____ + 19 =

_____ – 3 = → _____ ÷ 7 =

_____ cubed = → _____ + 11 =

_____ ÷ 5 = → **Finish 15**

Name **Date**

1. **a.** $9^2 + 9 =$ ________ **b.** $24 + 12^2 =$ ________ **c.** $15^2 - 25 =$ ________
   d. $2^3 + 2 =$ ________ **e.** $10^3 + 10^2 + 10 =$ ______ **f.** $10^3 - 10^2 =$ ________
   g. What triangular number is built from a base of 9? ________

2. Complete the factor tree below.

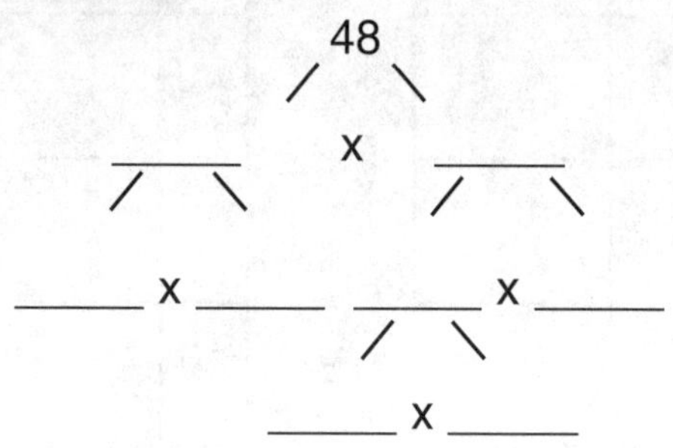

48 = ______________________

48 = ______________________

3. Complete the table. Follow the arrows to help you.

n	n x 8	n x 8 + 12
7		
	96	
		76
		60
25		

4. Place the following numbers in their correct columns. Some belong in more than one column. One does not belong in any column.

432, 126, 693, 245, 252, 824, 774, 516

Divisible by 9	Divisible by 4	Divisible by 6

5. **a.** sum of 16 and 14 = _____ **b.** total 23, 6, and 9 _____
   c. difference between 45 and 29 _____ **d.** 15 less than 64 = _____
   e. remainder when 19 is divided by 7 _____ **f.** quotient of 664 and 8 _____
   g. 6 and _____ are factors of 54 **h.** Is 85 a multiple of 5? _____
   i. 13 and _____ are addends of 30 **j.** $3^2 \times 2^3 =$ _____

NUMBER PATTERNS

Through work with number patterns, students learn that mathematics does not consist entirely of straight operational work with numbers and shapes. Skills of observation, generalization, visualization, and approximation must be developed.

In these units, patterns are explored through tests for divisibility, work with composites and primes, square and triangular numbers, and using arrays and multiples. Number patterns are completed and their rules are discovered and written. Missing numbers in a series are found and numbers that do not fit are identified. Patterns are used to solve problems. Two assessment pages are included.

UNDERSTANDING NUMBER PATTERNS

Unit 1

Patterns
Series
Prime and composite numbers
Square and triangular numbers
Multiples
Problems and puzzles

Objectives

- *generate and describe number patterns using a variety of strategies:*
  - * *prime and composite numbers*
  - * *square and triangular numbers*
  - * *tests of divisibility and multiples*
  - * *the four processes*
- *apply rules for patterns which relate elements to their position*
- *apply rules for patterns which relate elements to the previous element*

Language

series, sequence, prime, composite, square, triangular, <, >, multiple, divisible, positive, consecutive

Materials/Resources

colored pencils, calculators

Contents of Student Pages

* *Materials needed for each reproducible student page*

Remember

- ❑ *Encourage children to see the patterns in all numerical situations.*
- ❑ *Help parents to foster this in children.*

Additional Activities

- ❑ *Look for geometric patterns in the environment such as on buildings, floorings, and playground equipment.*
- ❑ *Look for patterns of numbers and letters on car license plates, on street numbers, and in phone numbers.*
- ❑ *Use pattern recognition as a method of solving problems.*
- ❑ *Use children's birthdays and initials to make patterns.*
- ❑ *Use numbers as foundations for designs.*

Answers

Page 96 Prime and Composite Numbers

1. Check individual work.
2. 2, 3, 5, 7, 11, 13, 17, 19, 23, 29, 31, 37, 41, 43, 47, 53, 59, 61, 67, 71, 73, 79, 83, 89, 97
3. 7, 17, 37, 47, 67, 97
4. 2, 3, 5, 7, 11, 13, 17, 19
5. False
6. 32, 33, 34, 35, 36, 38, 39

Page 97 Digit Patterns

1. a. 0
   b. 5 or 0
   c. 2, 4, 6, 8, or 0
   d. 3, 6, 9, 2, 5, 8, 1, 4, 7, 0
   e. 4, 8, 2, 6, 0
   f. all contain numbers 2, 4, 6, 8, 0
2. Check individual work.
3. a. even
   b. 72, 81, 54, 99, 42, 48
4. Yes. The last two digits in every number are divisible by 4.
5. multiples of 3: 54, 78, 84, 27, 39, 231, 93
   multiples of 4: 84, 136, 248, 212, 424

Page 98 Square Numbers

1. Check individual arrays.
   5 x 5 = 25, 4 x 4 = 16
2. 1, 4, 9, 16, 25, 36, 49
3. True
4. False
5. True
6. 3, 5, 7, 9, 11, 13, 15, 17, 19
7. the pattern of rising odd numbers
8. Squared minus one. (2, 3) (6, 35) (9, 80) (11, 120) (1, 0) (20, 399) (12, 143)
9. a. 225
   b. 625
   c. 2,500
   d. 9,801
   e. 3,969
   f. 2,304

Page 99 Triangular Numbers

1. a. 1 + 2 + 3 + 4 + 5 + 6 + 7 = 28
   b. 1 + 2 + 3 + 4 + 5 + 6 + 7 + 8 = 36
   c. 1 + 2 + 3 + 4 + 5 + 6 + 7 + 8 + 9 = 45
2. a. $16 = 4^2$
   b. $36 = 6^2$
   c. $64 = 8^2$
   d. $81 = 9^2$
   e. $49 = 7^2$
3. Answers will vary.
4. a. 646 R, 876 L, 702 R, 960 L, 222 L, 031 R, 541 L, 988 R

Page 100 Series and Number Patterns

1. a. 0.5, 1.5, 2.5, 3.5, 4.5
   b. 10, 7.5, 5.0, 2.5, 0
   c. 160, 80, 40, 20
2. Check individual work.
3. a. 60, 66
   b. 40, 32
   c. 32, 64
   d. 49, 64
   e. 6.0, 7.5
   f. 3, 1
4. a. 11, 2, 6, 29
   b. 16, 12, 28, 34
   c. 10, 36, 3, 56
5. B: 2, 0.5, 3, 5, 6,
   C: 48, 49, 2
6. High jump: 16, 24, 32, 40, 56, 88
   200-m race: 9, 14, 20, 30, 38, 54

Page 101 Problem Solving

1. HJL 818 (add 91 each time)
2. They live in triangular numbered houses on streets whose names use alliterations; No.
3. $22 (price increases $3 per watermelon)
4. $50 (discount: 1/2, 2/3, 3/4, 4/5, 5/6)
5. 8 laps; 16 laps 80 minutes
   (laps: double; minutes: +5, +10, +15, +20)

Page 102 Assessment

1. a. $196 (buy 2 or more at $48 each)
   b. 4 (prime numbers)
   c. 2nd house is #18, it has only one cross on antenna; #24 has no division in lower window; #30 has upper window on right side; #24 has different tree
   d. 125 (5^3)
2. Square: 9, 81, 49, 121
   Prime: 17, 23, 31, 11, 47, 2
   Multiples of 4: 44, 80, 124, 52, 68
   Triangular: 15, 21, 6, 55, 78
3. a. 13
   b. 28
   c. 81
   d. 36

Name **Date**

1	2	3	4	5	6	7	8	9	10
11	12	13	14	15	16	17	18	19	20
21	22	23	24	25	26	27	28	29	30
31	32	33	34	35	36	37	38	39	40
41	42	43	44	45	46	47	48	49	50
51	52	53	54	55	56	57	58	59	60
61	62	63	64	65	66	67	68	69	70
71	72	73	74	75	76	77	78	79	80
81	82	83	84	85	86	87	88	89	90
91	92	93	94	95	96	97	98	99	100

1. **a.** Underline ten, then cross every multiple of ten with a yellow colored pencil.
   b. Underline nine, then cross every multiple of nine with a red colored pencil.
   c. Underline eight, then cross every multiple of eight with a green colored pencil.
   d. Underline seven, then cross every multiple of seven with a blue colored pencil.
   e. Underline six, then cross every multiple of six with a pink colored pencil.
   f. Underline five, then cross every multiple of five with an orange colored pencil.
   g. Underline four, then cross every multiple of four with a purple colored pencil.
   h. Underline three, then cross every multiple of three with a gray colored pencil.
   i. Underline two, then cross every multiple of two with a black colored pencil.
2. Circle the numerals larger than 1 that are uncrossed. They are prime numbers. Write them. ______________________________
3. Write the prime numbers that end in seven. ______________________________
4. Write the prime numbers less than 20. ______________________________
5. All the prime numbers less than 20 are even. True or false? ____________________
6. The numerals that are crossed are composite numbers because they are multiples of numbers greater than 1. Write the composite numbers between 30 and 40.

____

Name **Date**

1. Multiples of a number have a pattern with the digits in the ones place. This is called the *digit pattern*.
   - a. Every multiple of 10 ends with ______________
   - b. Multiples of 5 end with __________ or __________
   - c. Multiples of 2 end with ______________________________
   - d. The digit pattern for 3 is ______________________________
   - e. The digit pattern for 4 is ______________________________
   - f. How are the digit patterns for 3 and 4 similar to the digit pattern for 2? __________

   ____

2. Connect the digits for the digit pattern given to show a design.

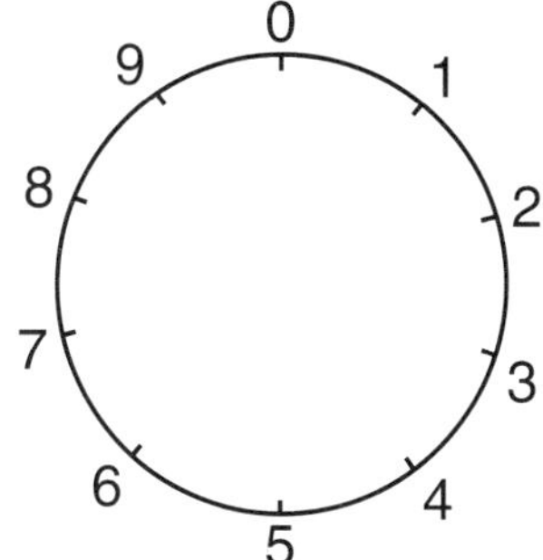

Digit pattern for 2

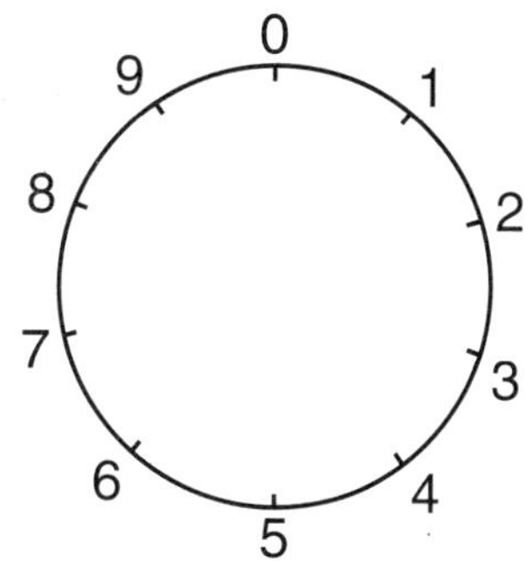

Digit pattern for 3

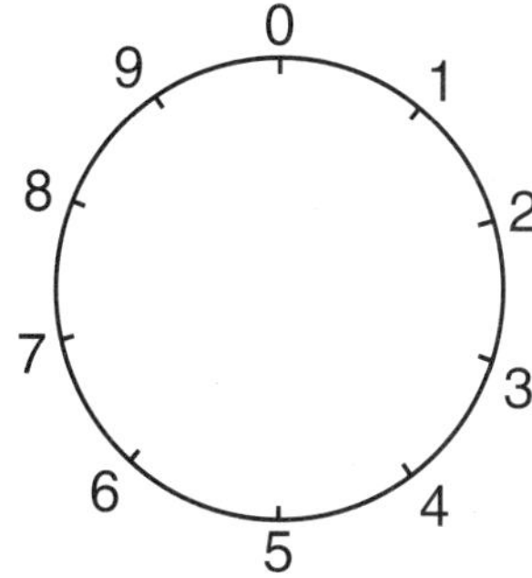

Digit pattern for 4

3. a. Numbers divisible by 2 are all ______________ numbers. They end in 2, 4, 6, 8, or 0.

> What about the numbers divisible by 3?
>
> 3, 6, 9, 12, 15, 18, 21 . . .
>
> The *sum* of the digits is always divisible by 3! Keep adding the sum of the digits until you get a single figure.

   b. Circle the numbers divisible by 3.

   17 72 81 89 54 64 99 83 42 48

4.

> A number is divisible by 4 if the last two digits are divisible by 4.
> Is 232 divisible by 4? Yes! 32 is divisible by 4.

Are all these numbers divisible by 4? 516, 120, 288, 124, 248, 312

Why? ____

5. Circle the multiples of 3 and underline the multiples of 4.

54 65 78 84 136 25 27 248 39 115 231 212 93 235 424

Name **Date**

- We can illustrate 3 x 3 = 9 using an array.
- Because this forms a square, we call 9 a square number.

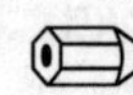 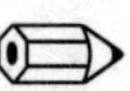 

1. Complete these arrays to show that 25 and 16 are square numbers.

_____ x _____ = 25 _____ x _____ = 16

2. Write the square numbers less than 50. ____________________

3. All square numbers except 1 are composite numbers. True or false? __________

4. All square numbers are even numbers. True or false? __________

5. The smallest positive square number is 1. True or false? __________

6. Write the difference between each pair of square numbers in the box between them.

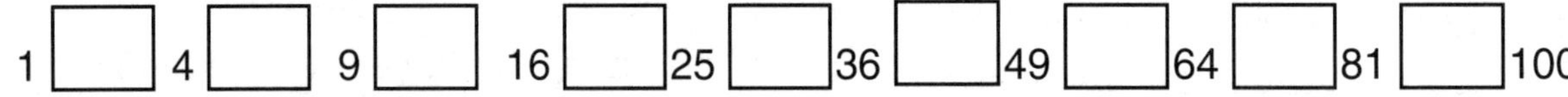

7. What pattern do these differences form? ____________________

____

8. What happened to each number below to make its partner?
Find out and calculate all the missing partners. __________

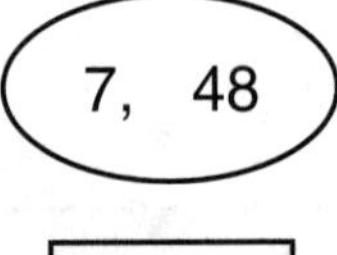

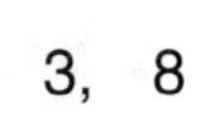

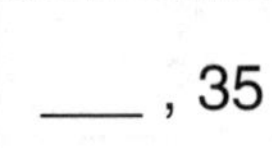

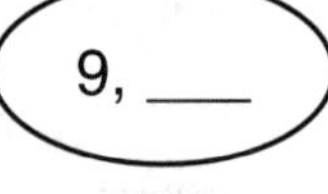

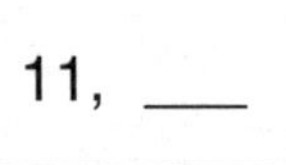

10,
99

___ , 143

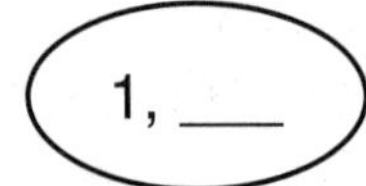

9. Use a calculator to find the square of:

a. 15 __________ **b.** 25 __________ **c.** 50 __________

d. 99 __________ **e.** 63 __________ **f.** 48 __________

Name	Date

Some numbers form a triangular pattern

1 ○
3 ○○
6 ○○○
10 ○○○○

3, 6, and 10 are triangular numbers.

This is the triangular number pattern.

1 + 2 + 3 + **4** = 10

1 + 2 + 3 + 4 + **5** = 15

1 + 2 + 3 + 4 + 5 + **6** = 21

1. Continue the triangular number pattern.

   a. 1 + 2 + 3 + 4 + 5 + 6 + 7 = ____________________

   b. ____________________

   c. ____________________

Square numbers are the sum of two consecutive triangular numbers. $1 + 3 = 4 = 2^2$

2. Triangular numbers.

   a. 6 + 10 = ______ = ______

   b. 15 + 21 = ______ = ______

   c. 28 + 36 = ______ = ______

   d. 36 + 45 = ______ = ______

   e. 21 + 28 = ______ = ______

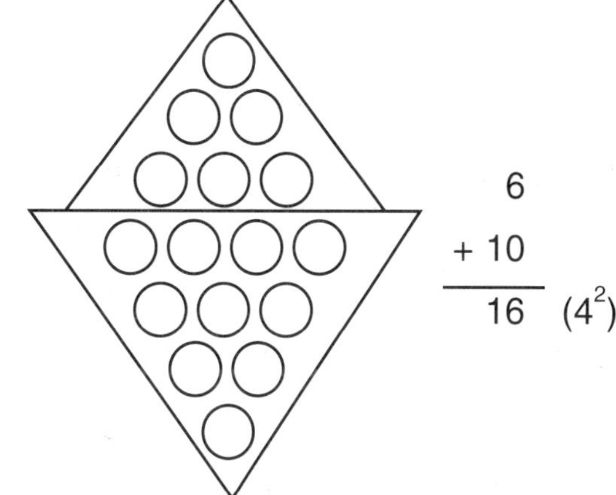

3. **a.** Divide a different square number into the sum of two consecutive triangular numbers, as above.

   b. Write the number sentence to describe this fact.

   $\_\_\_\_^2$ = _____ = _____ + _____

4. Cars with license-plate digits adding to square numbers must park to the right and cars with license-plate digits adding to triangular numbers must park to the left. Direct these cars, right (R) or left (L).

646	876	702	960	222	031	541	988

Name	**Date**

1. Write the following series as directed.
   a. Begin at 0.5 and add 1 each time. _____ _____ _____ _____ _____
   b. Begin at 10 and subtract 2.5 each time. _____ _____ _____ _____ _____
   c. Begin at 160 and halve the number each time. _____ _____ _____ _____
2. This number line shows the sequence (+5, –2). Continue it.

0 2 4 6 8 10 12 14 16 18 20 22 24 26

3. Study the following series and write the next two numbers.
   a. 36, 42, 48, 54, _____ , _____
   b. 72, 64, 56, 48, _____ , _____
   c. 1, 2, 4, 8, 16, _____ , _____
   d. 9, 16, 25, 36, _____ , _____
   e. 1.5, 3.0, 4.5, _____ , _____
   f. 81, 27, 9, _____ , _____
4. Follow the rule for each number pattern. Fill in the empty squares.

a. A = 3 x B – 1

A	B
	4
5	
17	
	10

b. H = J ÷ 2

H	J
8	
	24
	56
17	

c. X = Y x Z – Z

X	Y	Z
	3	5
	5	9
3	2	
	9	7

5. Each column is a separate vertical number pattern. Start at the top of each column and work your way down. The pattern for each column is the same: A x (B x 2) = C. Supply the missing numbers.

A	5	3	8	12	15	25	7	60	9	1
B	2	4		2			3.5			1
C	20	24	32		15	150		600	108	

6. At the Numbers Olympics, the competitors became muddled in the change rooms. The following clues will help you send them to the correct venue.
   All (✻ x 4) were to be at the high jump, while all (★ + 7) were to be at the 200-meter race. ✻ are all even numbers, ★ are all prime numbers.

9, 14, 16, 20, 24, 30, 32, 38, 40, 54, 56, 88

High Jump ✻	**200-meter Race ★**

Name **Date**

Look for a pattern as a strategy to solve problems

1. Joe's family has a tradition that the car license plate numerals and letters must follow a pattern that was started by their great grandfather in 1910. If these are the family's previous license plates, what will be the new one for Joe?

ACE 181	BDF 272	CEG 363	DFH 454
EGI 545	FHJ 636	GIK 727	← Joe's License Plate

2. From their addresses, you can tell that the McDonalds will only live on certain streets with special numbers. What is their secret?

   28 Sanderson St. : 6 Pine Place : 78 Albartos Ave. : 21 Cross Ct. : 105 Reily Rd.

   The McDonalds' secret is ______________________________

   Will they live in 93 Bennet Blvd.? ______________________________

3. Watermelon prices at the farmer's market are as follows:

   3 melons for $7 | 4 melons for $10 | 5 melons for $13

   What would they charge for 8 watermelons? __________

4. **Keep Fit Classes! Sign up today!**
   Regularly $60 per hour
   Every person gets a Mystery Discount

   1st person to sign up pays $30 per hour
   2nd person to sign up pays $40 per hour
   3rd person to sign up pays $45 per hour
   4th person to sign up pays $48 per hour

   At this rate, what will the 5th person to sign up pay per hour? __________

5. What numbers are missing in the Fitness Program? The fitness program advertised is:

   Monday: 1 lap plus 30 minutes on the exercise bike

   Tuesday: 2 laps plus 35 minutes on the bike

   Wednesday: 4 laps plus 45 minutes on the bike

   Thursday: ___ laps plus 1 hour on the bike

   Friday: ___ laps plus ___ minutes on the bike

Name **Date**

1. Visitors to Patternsville have a tough time working out the patterns on signs and buildings. At the Information Kiosk, you get questions like this. Can you help?

   a. What do 4 sleeping bags cost? _____

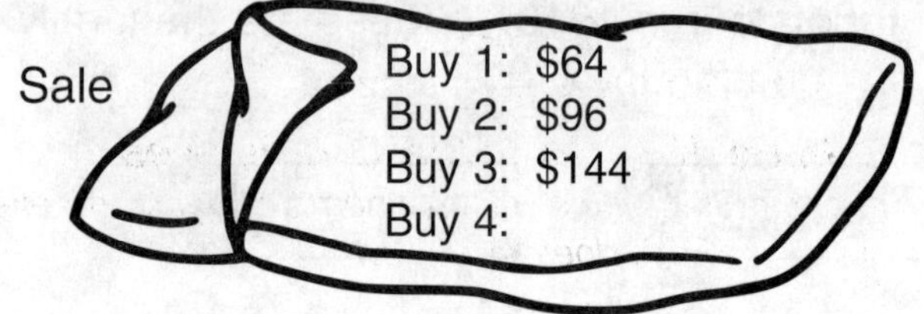

   b. What highway number has fallen off? _____

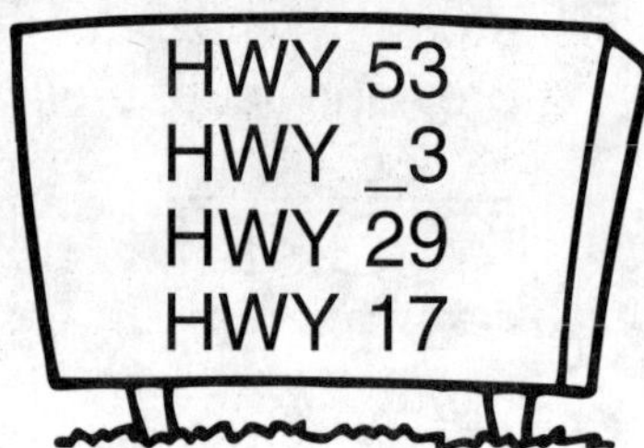

   c. What's wrong in this street? (5 things)

____

____

____

   d. What is the missing door number on the building?

   ____

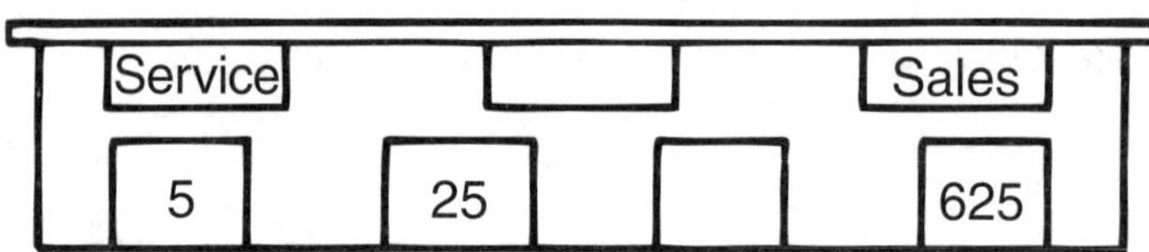

2. Professor Summup left his lab in a mess, and his assistant is left to sort out the numbers the professor has been working on. Please help put all the numbers in the correct bins.

   17, 9, 15, 23, 21, 44, 31, 81, 6, 80, 124, 55, 49, 47, 11, 121, 78, 52, 2, 68

Square

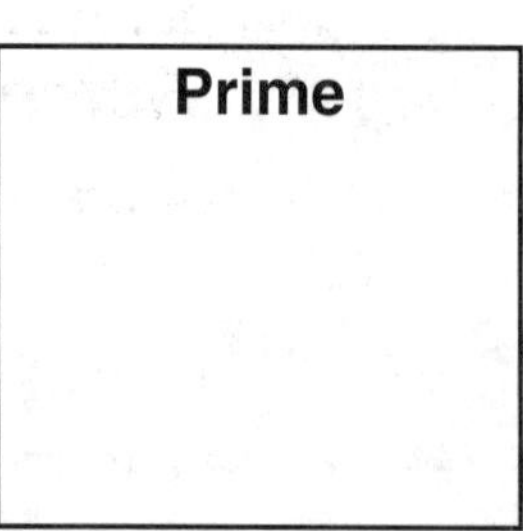

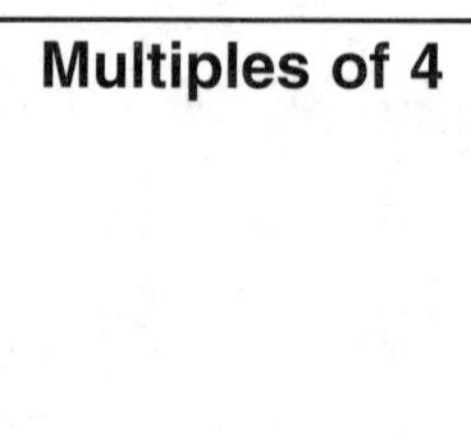

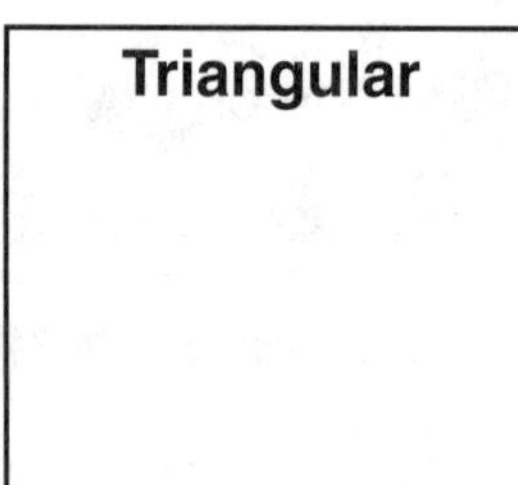

3. Find the missing number in each pattern.

   a. 5, 9, _____ , 17, 21

   b. 62, 52, 43, 35, _____

   c. 3, 9, 27, _____ , 243

   d. 72, _____ , 18, 9

WORKING WITH NUMBER PATTERNS

Unit 2

Series
Square and triangular numbers
Prime and composite numbers
Multiples and digit patterns
Problems
Puzzles

Objectives

- *generate and describe number patterns using a variety of strategies:*
  - *the four processes*
  - *square and triangular numbers*
  - *prime and composite numbers*
- *apply rules for patterns which relate elements to their position*
- *apply rules for patterns which relate elements to their previous element*

Language

series, sequence, prime, composite, square numbers, triangular numbers, multiple, divisible

Materials/Resources

highlighters

Contents of Student Pages

* *Materials needed for each reproducible student page*

Remember

- ❑ *Review Unit 1. All terms used there are built upon in Unit 2.*

Additional Activities

- *Encourage the search for patterns in everyday situations.*
- *Encourage the making of patterns.*
- *Use recognition of patterns as a means of solving puzzles and problems.*
- *Run trivia sessions on mathematical sequences, multiples patterns, and missing numbers.*
- *Run a daily puzzle that involves a pattern.*
- *Play games that foster the use of patterns.*
- *Look for patterns in nature.*

Answers

Page 105 Series

1. a. 67, 117, 167, 217, 267, 317
   b. 570, 520, 470, 420, 370, 320
   c. ↓ ← ↑ → ↓
   d. ⅃ M ᗺ ꓘ Ƨ
2. a. 42, 49, 63
   b. 63, 36, 18
   c. 25, 32, 49
   d. 53, 49, 38
   e. 27, 81, 729
   f. 10, 9, 18
   g. 1,000, 10,000
   h. 2.0, 2,000, 20,000
3. a. 2 x A + B = C
   b. J + (K ÷ 2) = L
   c. (M + N) ÷ 3 = O
   d. (X + Y) x 5 = Z
4. a. 0, 5, 3, 8, 6, 11, 9, 14, 12, 17, 15
   b. 25, 30, 45, 50, 65, 70, 85, 90, 105, 110, 125, 130
   c. 5, 25, 20, 40, 35, 55, 50
   d. 0, 5, 15, 30, 50, 75, 105, 140
   e. 0.1, 0.2, 0.4, 0,7, 0.8, 1.0, 1.3, 1.4, 1.6

Page 106 Patterns Using Multiples

1. a. decreasing by 1
   b. increasing by 1
   c. 9
2. 36, 234, 450, 297, 117, 144, 135, 81, 45
   b. X
3. a. 8, 16, 24, 32, 40, 48, 56, 64, 72, 80
   b. 8, 6, 4, 2, 0
   c. No
4. a. 21, 35, 42, 56, 63, 77
   b. 1, 8, 5, 2, 9, 6, 3, 0
   c. Check individual work.
   d. a star
   e. 14 and 28 are multiples of 7

Page 107 Codes

1. Take shelter until further notice.
2. Unable to find suitable hideaway.
3. Answers may vary. Any foods that contain double letters are correct (apples, pizza, carrots, etc.).
4. Meeting with Premier Train scheduled eight a.m. Take notes about Transport shortage.
5. HELP (Square numbers; numbers represent the placement of letters in the alphabet.)

Page 108 Odd One Out/Puzzles

1. a. 74—not multiple of 9
   b. 72—not square
   c. 27—not even
   d. 2—not odd
   e. 125—not multiple of 4
   f. hectare—not linear measure
   g. relish—no double letters
   h. 12—not triangular
   i. 2nd 8—not in digit pattern of 7
   j. 27—not prime
   k. 7—not double 4
   l. pyramid—not a quadrilateral
   m. 231—not having digit sum of 9
   n. 111—not a multiple of 11
   o. square—not 3D
2. stoon vun, stoon stoo, stoon stree, stoon tor, streen vun, streen stoo, streen stree, streen tor, torn vun, torn stoo, torn stree, torn tor
3. a. 4 + 3 + 2 + 1 + 0 = 10
   b. 5 + 4 + 3 + 2 + 1 + 0 = 15
   c. 6 + 5 + 4 + 3 + 2 + 1 + 0 = 21
   d. 7 + 6 + 5 + 4 + 3 + 2 + 1 + 0 = 28
   e. Triangular numbers

Page 109 Puzzles and Problems

1. 20 x 3 + 1 = 61
   50 x 3 + 1 = 151
2. 3 + (24 x 2) = 51
3. Phone home immediately
4. a. 4 x (4 + 1) = 20
   b. 20 x (20 + 1) = 420
5. a. 3 x 3 = 9
   b. 20

Page 110 Assessment

1. a. 144, 153
   b. 46, 51
   c. 3.9, 4.8
   d. 2,560, 20,480
   e. 14, 10
2. a. 3 + 5 + 7 + 9 + 11 + 13 = 49 b. 1 + 3 + 5 = 9
3. 1,248; 1,320; 2,640; 1,480; 3,400; 6,280; 7,240; 4,880
4. Lost North of Broome
5. Yes; it is a compound word.
6. a. 71—not a multiple of 3
   b. 80—not 82 + 2 (pattern is –4, +2)
   c. 200—not square
   d. lead—changed two letters, not one or
      step—no double vowel
   e. 55—not minus 20
7. 61
8. a. A
   b. C
   c. C
   d. A

Name **Date**

1. Complete each series according to the directions.
   a. Add 50. 17, _____ , _____ , _____ , _____ , _____ , _____
   b. Subtract 50. 620, _____ , _____ , _____ , _____ , _____ , _____
   c. Turn 90° clockwise. ➔ _____ , _____ , _____ , _____ , _____
   d. Mirror reverse. L _____ M _____ B _____ K _____ S _____

2. What are the missing numbers in these series?
   a. 21, 28, 35, ___ , ___ , 56, ___
   b. 72, ___ , 54, 45, ___ , 27, ___
   c. 7, 10, 14, 19, ___ , ___ , 40, ___
   d. 59, 58, 56, ___ , ___ 44, ___
   e. 1, 3, 9, ___ , ___ , 243, ___
   f. 2, 4, 3, 6, 5, ___ , ___ , ___ , 17
   g. 10, 100, ______ , ______ , 100,000
   h. 0.2, ___ , 20, 200, _____ , _____

3. Write the number sentence to describe the pattern in each set (e.g., A + B = C).

a.

A	B	C
2	5	9
3	7	13
2	10	14
0	4	4

____

b.

J	K	L
5	6	8
1	2	2
10	2	11
9	4	11

____

c.

M	N	O
6	3	3
12	6	6
8	13	7
20	16	12

____

d.

X	Y	Z
4	1	25
6	2	40
7	8	75
1	0	5

____

4. Continue these patterns on the number lines then write the series.
   a. 0 2 4 6 8 10 12 14 16 18
   ____
   b. 20 30 40 50 60 70 80 90 100 110 120 130
   ____
   c. 5 10 15 20 25 30 35 40 45 50 55
   ____
   d. 0 25 50 75 100 125 150
   ____
   e. 0.1 0.2 0.3 0.4 0.5 0.6 0.7 0.8 0.9 1.0 1.1 1.2 1.3 1.4 1.5 1.6
   ____

Name	Date

1. The following are all multiples of nine: 9, 18, 27, 36, 45, 54, 63, 72, 81, 90, 99, 108
   a. What is the pattern in the ones? ____________
   b. What is the pattern in the tens? ____________
   c. What is the sum of the digits in every multiple of nine above? ____________

2. **a.** Highlight the digits that are divisible by 9 (multiples of 9).

36	199	69	19	144
209	234	114	135	254
163	80	450	95	235
91	81	245	297	100
45	122	284	156	117

   b. What letter is formed by the highlighted numerals? ________

3. **a.** Write the multiples of 8 to 80.

   ________ ________

   ________ ________

   ________ ________

   ________ ________

   ________ ________

   b. What is the digit pattern for 8?

   ____

   c. Very large numbers are divisible by 8 if their last three digits are divisible by 8. Example: 67,480 is divisible by 8 because 480 is divisible by 8.

   Is 2,465 divisible by 8? ________

4. Write the missing multiples of 7 in this series.
   a. 7, 14, _____ , 28, _____ , _____ , 49 , _____ , _____ , 70, _____
   b. The digit pattern for 7 is: 7, 4, _____ , _____ , _____ , _____ , _____ , _____ , _____ , _____
   c. Join the digits which make up the digit pattern for 7 on the "clock" below.

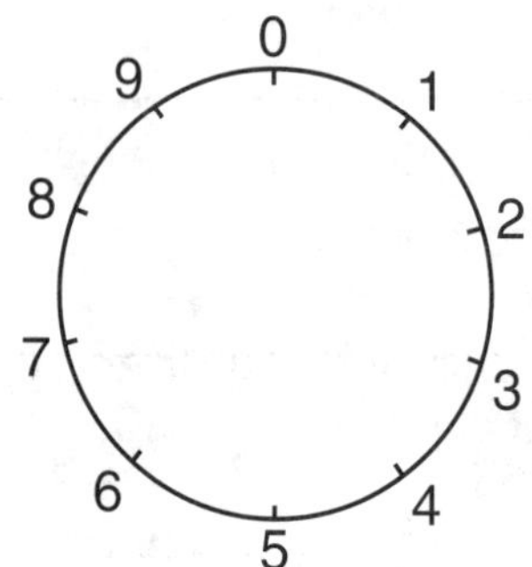

   d. What pattern is made? ____________

   ____

   e. The decimal equivalent for $\frac{1}{7}$ is 0.**1428**57.

   What do you notice about the bold digits in that decimal?

   ____

Name **Date**

Codes are made using patterns of numbers and letters. Good "code crackers" become very good at noticing patterns. Study these codes, complete them, and decode the messages.

1.

	20	21										
A	B	C	D	E	F	G	H	I	J	K	L	M
								14	15			
N	O	P	Q	R	S	T	U	V	W	X	Y	Z

Message: 12 19 3 23 11 26 23 4 12 23 10 13 6 12 1 4 24 13 10 12 26 23 10 6 7 12 1 21 23

____

2.

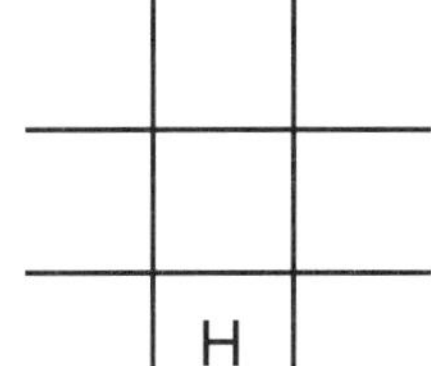

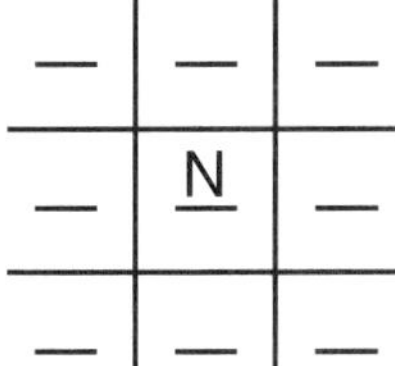

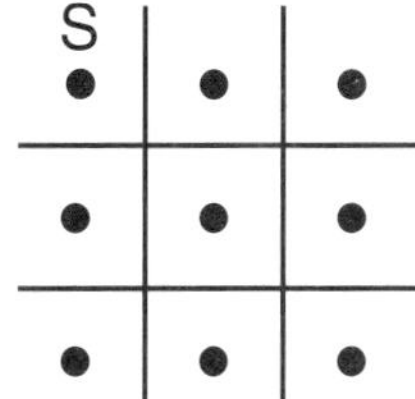

Message:

Decode: ____

3. When my pony was ill, he developed a liking for some very strange foods. I discovered he loved lettuce, but he hated celery. He loved cabbage, but couldn't stand cucumber. He happily chewed grass as well as rubber. "Would he like jelly?", I wondered. What other foods does my pony like? What is the pattern of his likes? (Hint: Look at the words, not their meaning.)

____

4. Ms. Blabb's computer keyboard has a case of "Viravowel," and her message to her boss reads like this:

Mwwtzng wzth Prwmzwr Trxzn schwdklwd wzght xm. Txkw nqtws xbqkt Trxnspqrt shqrtxgw.

What did she intend to say? ____

____

5. Shipwrecked marines wrote this message in the sand, knowing that their searchers would be able to decode it. Can you?

64 25 144 256

____

Name **Date**

1. Highlight the odd one out in each group of numbers or words and explain your choice.
   - **a.** 27, 36, 18, 45, 63, 74, 81 ____________
   - **b.** 100, 81, 64, 72, 49, 9 ____________
   - **c.** 20, 22, 24, 26, 27, 28, 30 ____________
   - **d.** 2, 3, 5, 7, 9, 11, 13 ____________
   - **e.** 64, 88, 112, 125, 136, 168, 160 ____________
   - **f.** centimeter, meter, hectare, kilometer ____________
   - **g.** apple, spaghetti, relish, pizza, cheese ____________
   - **h.** 36, 28, 21, 15, 12, 10, 6, 3, 1 ____________
   - **i.** 7, 4, 1, 8, 5, 8, 2, 9, 6, 3, 0 ____________
   - **j.** 7, 17, 27, 37, 47 ____________
   - **k.** (2, 4), (3, 6), (4, 7), (5, 10) ____________
   - **l.** pyramid, parallelogram, rhombus, rectangle ____________
   - **m.** 90, 135, 414, 153, 315, 231 ____________
   - **n.** 77, 55, 111, 88, 132, 121, 110 ____________
   - **o.** square, sphere, cube, prism, pyramid ____________

2. The Bundi people count using only five fingers on one hand, so they count like this: vun, stoo, stree, tor, jive, jive vun, jive stoo, jive stree, jive tor, stoon,...

   Can you keep counting from stoon, through to streen and torn to jivett?

   stoon, ____________ , ____________ , ____________ , ____________

   streen, ____________ , ____________ , ____________ , ____________

   torn, ____________ , ____________ , ____________ , ____________ , jivett

3. If every player shakes hands with every other player in a team of four, how many handshakes are made altogether?

   Player A shakes hands with 3 others: 3 handshakes

   Player B shakes hands with 3 others, but one (with A) has been counted: 2 handshakes

   Player C shakes hands with 3 others, but two (A and B) have been counted: 1 handshake

   Player D shakes hands with 3 others, but three (A, B, C) have been counted: 0 handshakes

   The pattern of handshakes for 4 people is 3 + 2 + 1 + 0 = 6
   - **a.** What is the pattern of handshakes for 5 people? ____________
   - **b.** What is the pattern of handshakes for 6 people? ____________
   - **c.** What is the pattern of handshakes for 7 people? ____________
   - **d.** What is the pattern of handshakes for 8 people? ____________
   - **e.** What type of numbers are all of these answers? ____________

Name **Date**

1. In the series 4, 7, 10, 13, the first number (4) equals 1 x 3 + 1, the second number (7) equals 2 x 3 + 1 and the third number (10) equals 3 x 3 + 1. What will be the twentieth and fiftieth numbers in this series? 20th = __________ 50th = __________

2. It takes three sticks to outline the first triangle in this pattern and only two to build every following triangle. Two triangles will take (3 + 2) or 3 + (1 x 2) sticks, three triangles will take 3 + (2 x 2) sticks, and 4 triangles will take 3 + (3 x 2) sticks. How many sticks will be needed to make twenty-five triangles?__________________

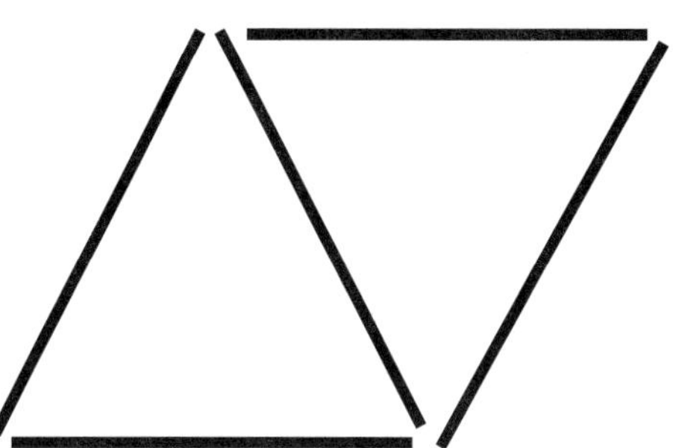

3. Complete the code, then decode the message.

A	B	C	D	E	F	G	H	I	J	K	L	M
*				*		5		*				10
N	O	P	Q	R	S	T	U	V	W	X	Y	Z
	*						*				20	

Message: 12, 6, *, 11, * 6, *, 10, * *, 10, 10, *, 3, *, *, 16, *, 9, 20

____

4. The sum of even numbers.

1st even number = 2 Sum of the first two = 6 ⟶ 2 x (2 + 1) = 6

2nd even number = 4 Sum of the first three = 12 ⟶ 3 x (3 + 1) = 12

3rd even number = 6 Sum of the first four = 20 ⟶ **a.** ______ x ______ = ______

4th even number = 8

b. What is the sum of the first 20 even numbers? ______________________

5. The sum of odd numbers.

1st odd number = 1 Sum of the first two = 4 ⟶ 2 x 2 = 4

2nd odd number = 3 Sum of the first three = 9 ⟶ **a.** _____ x ______ = ______

3rd odd number = 5

b. How many odd numbers are in the group whose sum is 400? ______________

Name **Date**

1. Fill in the missing numbers in each pattern.
   a. 108, 117, 126, 135, _____ , _____
   b. 31, 36, 41, _____ , _____
   c. 1.2, 2.1, 3.0 _____ , _____
   d. 5, 40, 320, _____ , _____
   e. 18, 14, 16, 12, _____ , _____

2. Complete to show that a square number is the sum of consecutive odd numbers.
   a. 7^2= 1 + _____ + _____ + _____ + _____ + _____ + _____ = _____
   b. 3^2 = _____ + _____ + _____ = _____

3. Travel only on the numbers that are multiples of eight. Shade in your path.

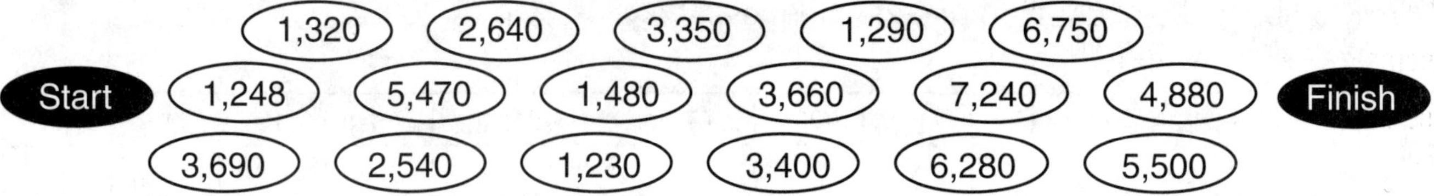

4. Please help Agent 007 complete this decoded message.

   15, 12, 8, 7 13, 12, 9, 7, 19 12, 21 25, 9, 12, 12, 14, 22

   L O S T ____

5. When you come to my party, you can bring watermelon, but please don't bring cake. I'd love you to bring pineapple, but not apples. I'm happy for you to bring newspapers, but not magazines. Will you come to my party bringing cupcakes? ____________

   Why? ____

6. Circle the odd one out and give a reason for your choice.
   a. 51, 54, 63, 69, 71, 75______________________________
   b. 88, 84, 86, 82, 80______________________________
   c. 100, 200, 400, 900, 1,000 ______________________________
   d. step, seep, seed, deed, lead ______________________________
   e. 100, 95, 85, 70, 55, 25______________________________

7. How many sticks are needed to complete 20 squares?______________________

8. Circle the shape that will complete each pattern.
   a. A B C D
   b. A B C D
   c. A B C D
   d. A B C D

Skills Index

The following index lists specific objectives for the student pages of each unit in the book. The objectives are grouped according to the sections listed in the Table of Contents. Use the Skills Index as a resource for identifying the units and student pages you wish to use.

Skills Index